PRINCIPLES OF CEREAL SCIENCE AND TECHNOLOGY

SECOND EDITION

R. Carl Hoseney

Department of Grain Science and Industry
Kansas State University
Manhattan, Kansas

Published by the
American Association of Cereal Chemists, Inc.
St. Paul, Minnesota, USA

Library of Congress Catalog Card Number: 93-74154
International Standard Book Number: 0-913250-79-1

©1986, 1994 by the American Association of Cereal Chemists, Inc.

All rights reserved.
No portion of this book may be reproduced in any form, including photocopy, microfilm, information storage and retrieval system, computer database or software, or by any means, including electronic or mechanical, without written permission from the publisher.

Reference in this publication to a trademark, proprietary product, or company name is intended for explicit description only and does not imply approval or recommendation of the product to the exclusion of others that may be suitable.

Printed in the United States of America on acid-free paper

American Association of Cereal Chemists, Inc.
3340 Pilot Knob Road
St. Paul, MN 55121-2097, USA

Preface to the Second Edition

In the eight years or so since the first edition of this book appeared, the field of cereal chemistry has continued to grow and, I am afraid, also has become more complicated. Few people in the field were studying the rheology of doughs in the early 1980s, but in this edition it merits a chapter. The same is true of glass transitions and their tremendous influence on cereals and their products. Also, in this edition, the chapter on proteins has been split into two chapters, with one concentrating on the gluten proteins, an area in which advances are rapidly being made. The other new chapter in this edition is a short one on feed, as the feeding of animals consumes large quantities of cereal products. Additions and corrections were made to almost all the previous chapters. I would like to thank those who pointed out our errors, both of omission and commission, in the first edition and hope that you will do so with this edition.

The objective of this edition stays the same as it was previously, that is, to form a base for your study of cereal science. It is hoped that with this background you can go to the standard reference works in cereals. I have had people tell me that the first edition was not a good reference work. I agree; it was not intended to be. I also hope that readers learn that the study of cereals is fun and that it remains fun for them, as it has for me for the last 35 years.

I would like to thank the many people who allowed me to use their figures and data. They are acknowledged in the credits section at the end of the book. I also would like to thank Deirdre Bath and LaVonne Boetel for their help in getting this edition together and Jon Faubion, Huifen He, and Arlene Hamamoto for critically reading all or parts of the manuscript.

R. Carl Hoseney

Preface to the First Edition

The field of cereal science and technology is very broad and complicated. Cereals are complex biochemical entities that vary in composition and properties from year to year, from location to location, and from one cultivar to another. Cereal science is also complicated by the fact that the same raw material may be used to make different products. Therefore, the definition of "good quality" for a cereal such as wheat changes depending upon whether the wheat is used to make a loaf of bread or a cookie.

To understand cereals and their processing into products, the cereal scientist must be a jack-of-all-trades. He or she must understand chemistry (all areas), biochemistry, physics, engineering, and many other sciences. In addition, the scientist must also understand the "art" of the cookie baker and the brewmaster. Clearly, no one will accomplish all of these. Those of us who have mellowed with age in the field have long since reconciled ourselves to the fact that we cannot understand it all. We have forgotten how difficult this is for the new, bright, idealistic student to accept.

The purpose of this book is to provide such students with a basic background, something that they can stand on as they start their studies of how cereals work. In an attempt to accomplish that goal, the book is written as a text, not simply as a reference book. I hope that it will provide the student with the necessary background to understand the reference books listed at the end of the chapters. I also hope that the students using this book will find, as I have, that the study of cereal science is fun!

Many people allowed me to use their figures and data or gave permission for figures and tables to be reprinted. In the interests of simplicity, the credits for the figures are given in a section at the end of the book.

I would also like to thank many other people for their help with the book, particularly those students who have sat through my many hours of lecture and asked the key questions that made me realize that I was not explaining something clearly. I thank Kathy Zeleznak and Art Davis for their help with the figures, Judy Toburen for her typing, and Walter Bushuk and Y. Pomeranz for reviewing the manuscript.

R. Carl Hoseney

Contents

CHAPTER ONE
Structure of Cereals, 1
Wheat • Corn • Rice • Barley • Rye • Oats • Sorghum •
Pearl Millet • Triticale • Review Questions • Suggested Reading

CHAPTER TWO
Starch, 29
The Starch Granule • Chemical Composition •
Organization of the Starch Granule • Starches from Different Cereals •
Heating Starch in Water • Gelling and Retrogradation •
Heating Starch in Limited Water • Modified Starches •
Conversion of Starch to Sweeteners • Review Questions • Suggested Reading

CHAPTER THREE
Proteins of Cereals, 65
Structure • Classification of Proteins •
Properties of the Protein Solubility Groups • Variation in Protein Content •
Wheat Proteins • Proteins in Other Cereals • Review Questions •
Suggested Reading

CHAPTER FOUR
Minor Constituents of Cereals, 81
Nonstarchy Polysaccharides • Sugars and Oligosaccharides • Lipids •
Enzymes • Vitamins and Minerals • Review Questions • Suggested Reading

CHAPTER FIVE
Storage of Cereals, 103
Basic Types of Storage • Moisture, No. 1 for Safe Storage • Drying of Cereals • Aeration • Functional Changes and Indices of Deterioration • Microflora and Mycotoxins • Insects • Rodents • Review Questions • Suggested Reading

CHAPTER SIX
Dry Milling of Cereals, 125
The Milling Process • Products and Yield • Processing of Flour • Milling of Grains Other than Wheat • Review Questions • Suggested Reading

CHAPTER SEVEN
Wet Milling: Production of Starch, Oil, and Protein, 147
Corn • Wheat • Rice • Production of Oil from Cereals • Review Questions • Suggested Reading

CHAPTER EIGHT
Rice, Oat, and Barley Processing, 159
Rice • Oat Milling • Barley Pearling • Review Questions • Suggested Reading

CHAPTER NINE
Malting and Brewing, 177
Dormancy and After-Ripening • The Malting Process • Brewing • The Brewing Process • Distilled Products • Review Questions • Suggested Reading

CHAPTER TEN
Gluten Proteins, 197
Physical Properties of Gluten • Thermal Properties of Gluten • Genetics of Gluten • Synthesis of Gluten • Large Gluten Aggregates • Individual Proteins or Polypeptides • Commercial Isolation of Gluten • Uses of Wheat Gluten • Nutritional Quality • Review Questions • Suggested Reading

CHAPTER ELEVEN
Rheology of Doughs and Batters, 213
Rheology • Dynamic Rheological Measurements • Other Rheological Measurements • Doughs • Batters • Conclusions • Review Questions • Suggested Reading

CHAPTER TWELVE
Yeast-Leavened Products, 229
Bread-Making Systems • Dough Formation • Fermentation • Molding, Proofing, and Baking • Transformation of Doughs into Baked Products • Retrogradation and Staling • Other Types of Leavened Products • Flour Quality for Breadmaking • Review Questions • Suggested Reading

CHAPTER THIRTEEN
Soft Wheat Products, 275
Soft vs. Hard Wheat Flours • Chemical Leavening • Cookies • Cookie Flour Quality • What Happens During Cookie Baking • Crackers • Layer Cakes • Other Types of Cakes • Biscuits • Review Questions • Suggested Reading

CHAPTER FOURTEEN
Glass Transition and Its Role in Cereals, 307
What Is a Glass Transition? • What Causes Glass Transitions? • Factors That Affect the Glass Transition Temperature • Measurement of Glass Transitions • Effect of Plasticizer (Water) on Glass Transitions • Glass Transitions in Cereals • Importance of Glass Transitions in Cereal Products • Glass Transitions of Sugar Solutions • Review Questions • Suggested Reading

CHAPTER FIFTEEN
Pasta and Noodles, 321
Pasta • The Production Process • Noodles • Flour for Noodles • Noodle Making • Review Questions • Suggested Reading

CHAPTER SIXTEEN
Breakfast Cereals, 335
Cereals That Require Cooking • Ready-To-Eat Cereals • Review Questions • Suggested Reading

CHAPTER SEVENTEEN
Snack Foods, 345

Corn Products • Masa and Its Products •
Cohesive Properties of Nonwheat (Nongluten) Doughs • Synthetic Nuts •
Pretzels • Bagels • Review Questions • Suggested Reading

CHAPTER EIGHTEEN
Feeds, 359

Grinding • Pelleting • Feed Formulation • Alternatives to Grinding •
Specialty Feeds • Review Questions • Suggested Reading

Figure Credits, 367

Index, 373

PRINCIPLES OF
CEREAL
SCIENCE AND TECHNOLOGY
SECOND EDITION

Structure of Cereals

CHAPTER 1

Members of the grass family (Gramineae), which include the cereal grains, produce dry, one-seeded fruits. This type of fruit is, strictly speaking, a caryopsis but is commonly called a *kernel* or *grain*. The *caryopsis* (Fig. 1) consists of a fruit coat or *pericarp*, which surrounds the seed and adheres tightly to a *seed coat*. The seed consists of an *embryo* or *germ* and an *endosperm* enclosed by a *nucellar epidermis* and a seed coat. In general, all cereal grains have these same parts in approximately the same relationship to each other.

The chemical constituents of cereal grains are often compartmentalized. This plays a large role in grain storage stability. Grains often contain degradation enzymes and their substrates. Certainly if the two come in contact, the degradation process can easily start. However, if the enzyme or the substrate is protected from coming in contact with the other, the system is stable. An example of compartmentalization is that the lipids of wheat occur in deposits that are protected from the enzymes in the rest of the endosperm (Fig. 2).

The caryopsis of all cereals develops within floral envelopes, which are actually modified leaves. These are called the chaffy parts or *glumes*. In rice and most cultivars of barley and oats, the floral envelopes cover the caryopsis so closely and completely that they remain attached to the caryopsis when the grain is threshed and constitute the *hull* of those grains. In wheat, rye, corn, sorghum, and pearl millet, the grain and hull separate readily during threshing, and the grains are said to be naked because they have an uncovered caryopsis.

Wheat

A caryopsis or kernel of wheat is diagrammatically shown in both longitudinal and cross section in Fig. 3. The kernels average about

8 mm in length and weigh about 35 mg. The size of the kernels varies widely depending upon the cultivar and their location in the wheat head or spike. Wheat kernels are rounded on the dorsal side (the same side as the germ) and have a longitudinal crease the length of the ventral side (opposite the germ). The crease, which runs nearly the entire length of the kernel, extends nearly to its center. The two cheeks may touch and thus mask the depth of the crease. The crease not only makes it difficult for the miller to separate the bran from the endosperm with a good yield but also forms a hiding place for microorganisms and dust.

Wheat kernels vary widely in endosperm texture (hardness) and color.

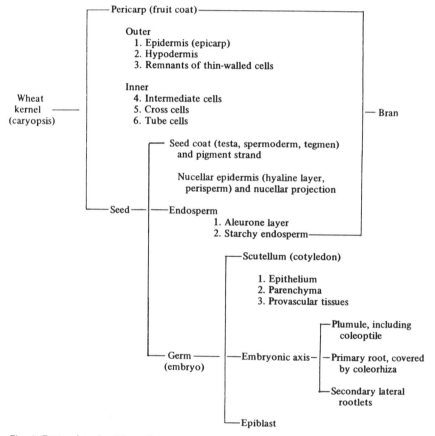

Fig. 1. Parts of a wheat kernel.

The variation in texture, which appears to be related to binding forces in the endosperm, is discussed later. The color, usually white or red (although purple is also known), is related to pigment in the seed coat. The type and presence of the pigments is under genetic control and thus can be manipulated by the plant breeder to give the desired color.

PERICARP

The pericarp surrounds the entire seed and is composed of several layers (Fig. 4). The outer pericarp is what millers call the beeswing.

Fig. 2. An aleurone cell showing the compartmentalization in cereals.

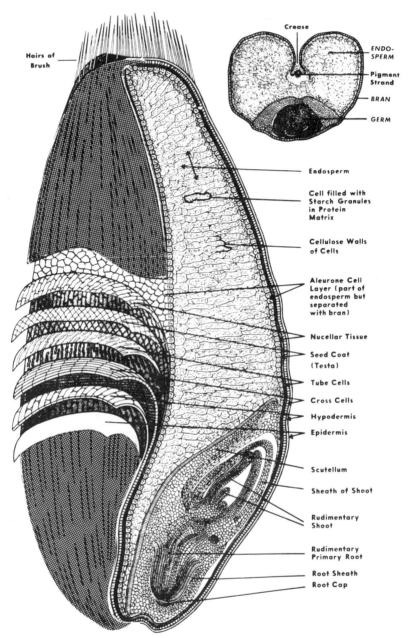

Fig. 3. Longitudinal and cross sections of a wheat kernel.

The innermost portion of the outer pericarp consists of the remnants of thin-walled cells; because of their lack of continuous cellular structure, they form a natural plane of cleavage. When they are disrupted, the beeswing is released. Removal of these layers also aids the movement of water into the pericarp.

The inner pericarp is composed of intermediate cells, *cross cells*, and *tube cells*. Neither the intermediate nor the tube cells completely cover the kernel. The cross cells are long and cylindrical (about 125 × 20 μm) and have their long axis perpendicular to the long axis of the kernel. The cross cells are tightly packed, with little or no intercellular space. The tube cells are of the same general size and shape as the cross cells but have their long axis parallel to the long axis of the kernel. They are not packed tightly and thus have many intercellular spaces. The total pericarp has been reported to comprise about 5% of the kernel and to consist of approximately 6% protein, 2.0% ash, 20% cellulose, and 0.5% fat, with the remainder being nonstarch polysaccharides.

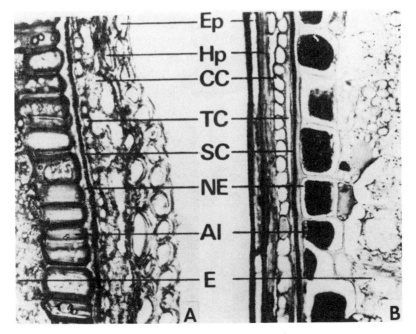

Fig. 4. Cross section (A) and longitudinal section (B) through the pericarp and adjacent tissues of a wheat kernel. Ep, epidermis; Hp, hypodermis; CC, cross cell; TC, tube cell; SC, seed coat; NE, nucellar epidermis; Al, aleurone layer; E, starchy endosperm.

SEED COAT AND NUCELLAR EPIDERMIS

The seed coat is firmly joined to the tube cells on the distal (outer) side and the nucellar epidermis on the proximal (inner) side. It consists of three layers: a thick outer cuticle, a layer that contains pigment (for colored wheats), and a thin inner cuticle. The seed coat of white wheat has two compressed cell layers of cellulose containing little or no pigment. The thickness of the seed coat varies from 5 to 8 μm. The nucellar epidermis, or hyaline layer, is about 7 μm thick and is closely united to both the seed coat and the aleurone layer.

ALEURONE LAYER

The *aleurone layer*, which is generally one cell thick, completely surrounds the kernel, covering both the *starchy endosperm* and the germ. From a botanical standpoint, it is the outermost layer of the endosperm. However, it is removed during milling, along with the nucellar epidermis, seed coat, and pericarp to form what the miller calls *bran*. The aleurone cells are thick-walled, essentially cuboidal, and free of starch at maturity (Fig. 5). The average size of the cells is about 50 μm. The cell walls are 3–4 μm thick and have been reported to be largely cellulosic in composition. Aleurone cells contain a large nucleus and a large number of *aleurone granules* (Fig. 5). The structure and composition of the aleurone granules is complex. The aleurone layer is relatively high in ash, protein, total phosphorus, phytate phosphorus,

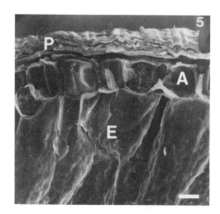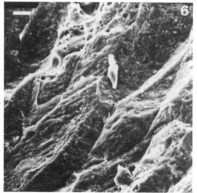

Figs. 5 and 6. Scanning electron micrographs of a cross section of a hard winter wheat kernel. Each bar is 20 μm.
  5. Pericarp (P), aleurone layer (A), endosperm (E).
  6. Endosperm cells.

fat, and niacin. In addition, thiamin and riboflavin are higher in the aleurone than in the other parts of the bran, and enzyme activity is high. Over the embryo, the aleurone cells are modified, becoming thin-walled cells that may not contain aleurone granules. The thickness of the aleurone layer over the embryo averages about 13 μm, or less than one third the thickness found elsewhere.

GERM OR EMBRYO

The germ of wheat comprises 2.5–3.5% of the kernel. As detailed in Fig. 3, the germ is composed of two major parts, the *embryonic axis* (rudimentary root and shoot) and the *scutellum*, which functions as a storage organ. The germ is relatively high in protein (25%), sugar (18%), oil (16% of the embryonic axis and 32% of the scutellum are oil), and ash (5%). It contains no starch but is rather high in B vitamins and contains many enzymes. The germ is quite high in vitamin E (total tocopherol), with values ranging up to 500 ppm. The sugars are mainly sucrose and raffinose.

ENDOSPERM

The starchy endosperm, excluding the aleurone layer, is composed of three types of cells: peripheral, prismatic, and central. The cells vary in size, shape, and location within the kernel. The peripheral cells are the first row of cells inside the aleurone layer and are usually small, being equal in diameter in all directions or slightly elongated toward the center of the kernel (Fig. 5). Several rows of elongated prismatic cells are found interior of the peripheral cells (Fig. 6). They extend inward to about the center of the cheeks and are about 150 × 50 × 50 μm in size. The central cells are interior of the prismatic cells. They are more irregular in size and shape than are the other cells.

The endosperm cell walls are composed of pentosans, other hemicelluloses, and β-glucans but not cellulose. The thickness of the cell walls varies with location in the kernel; they are thicker near the aleurone. Cell wall thickness also appears to vary among cultivars and between hard and soft wheat types (Figs. 7 and 8). The difference between *hard* and *soft* wheats may be the result of selection; hard wheats (bread wheats) have been selected for high water absorption. The hemicellulose in them absorbs large amounts of water. Thus, we are actually selecting for thick cell walls. In contrast, we do not want soft wheat flour to absorb large amounts of water, and thus we select for low water absorption (and thin cell walls).

Another difference between hard (Fig. 7) and soft (Fig. 8) wheats is the point of fracture when the kernels are broken. In hard wheat kernels, the first point of fracture occurs at the cell wall rather than through the cell contents. This is particularly evident in those cells just below the aleurone. In soft wheat endosperm, the fracture occurs primarily through the cell contents. This is evidence that the cell contents are more firmly bound to each other in hard wheats, resulting in a point of relative weakness between the cell walls. Of course, as the endosperm is reduced to flour size, the hard wheat cell contents are also fractured.

When reduced to appropriate particle size, the contents and cell walls of the endosperm cells make up *flour*. The cells are packed with starch granules embedded in a *protein matrix* (Fig. 9). The protein is mostly, but not entirely, *gluten*, the storage protein of wheat. In maturing wheat, the gluten is synthesized and deposited as *protein bodies*. However, as the grain matures, the protein bodies are compressed together into a matrix that appears mud- or claylike, and the protein bodies are no longer discernible. The starch granules occur as large, lenticular (lens-shaped) granules of up to 40 μm across the flattened side and as small spherical granules (2–8 μm in diameter). In actuality, one can find granules of all sizes between these extremes, but these two sizes and shapes are preponderant.

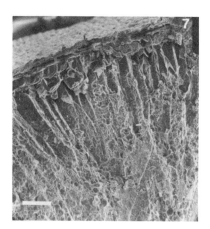 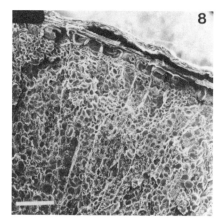

Figs. 7 and 8. Scanning electron micrographs (low magnification) of cross sections of wheat. Each bar is 100 μm.
   7. Hard winter wheat with breakage at the cell walls.
   8. Soft winter wheat with breakage through the cells.

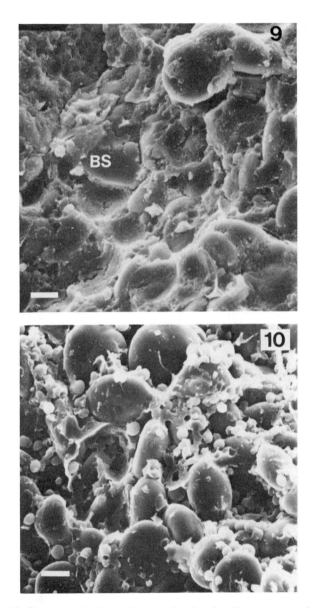

Figs. 9 and 10. Scanning electron micrographs showing the contents of endosperm cells. Each bar is 10 μm.
9. Hard winter wheat. Note the broken starch (BS) granule.
10. Soft winter wheat.

Figure 9, which is of hard wheat, also shows the tight adherence of the protein and starch. The protein appears to wet (coat or adhere to) the starch surface very well. This is characteristic of hard wheats. Not only does the protein wet the starch well, but the bond between the two is strong. Evidence for the strength of the bond is the tendency of hard wheats to break at the cell wall rather than through the cell contents and the breaking through some starch granules (note the broken granule in Fig. 9) rather than at the starch-protein interface.

In a similar micrograph of soft wheat (Fig. 10), the appearance is quite different. The starch and protein are similar in appearance to those of hard wheat, but the protein does not wet the surface of the starch. No starch granules are fractured, as the bond between the protein and starch ruptures easily, showing that it is not strong.

In durum wheat, which is much harder than the common hard wheats, a much larger number of broken starch granules occur when the kernel is fractured (Fig. 11). When sufficient force is applied across an endosperm cell, starch granules break, not the starch-protein bond. The strength of the protein-starch bond appears to explain kernel hardness. In soft wheats, the protein-starch bond ruptures easily, and

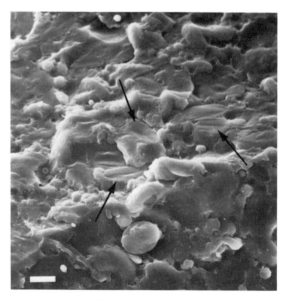

Fig. 11. Scanning electron micrograph of a durum wheat kernel, showing the contents of an endosperm cell. Note the large number of broken starch granules (arrows). Bar is 10 μm.

the kernel crushes with minimal force. In progressively harder wheats, the protein-starch bond is progressively stronger.

The nature of the starch-protein bond is not known. However, the fact that protein and starch can be easily separated from each other after treatment of flour with water seems to indicate that the bond is broken or weakened by water. Recent work has shown that the starch isolated from soft wheat cultivars contains a protein with a molecular weight of about 15,000 that is absent, or at least present at a much lower amount, in hard wheats. It has been suggested that the protein coats the starch and interferes with a strong protein-starch interaction in the endosperm. This presumably causes the grain to be soft.

In addition to the differences in hardness, another important characteristic of the wheat endosperm is its appearance. Some wheats are *vitreous*, hornlike, or translucent in appearance, while others are *opaque*, mealy, or floury. Traditionally, vitreousness has been associated with hardness and high protein content and opacity with softness and low protein. However, vitreousness and hardness are not the result of the same fundamental cause, and it is entirely possible to have hard wheats that are opaque and soft wheats that are vitreous, although these are somewhat unusual.

The air spaces in the kernel diffract and diffuse light and make the kernel appear opaque or floury. In tightly packed kernels, with no air space, light is diffracted at the air-grain interface but then travels through the grain without being diffracted again and again. The result is a translucent or vitreous kernel. As expected, the presence of air spaces within the grain makes the opaque grain less dense. The air spaces are apparently formed during the drying of the grain. As the grain loses water, the protein shrinks, ruptures, and leaves air spaces. With vitreous endosperm, the protein shrinks but remains intact, giving a denser kernel. If grain is harvested before it matures and is dried by freeze-drying, it is opaque. This shows that the vitreous character results during final drying in the field. It is also well known that vitreous grain that is wet and dried in the field, or for that matter in the laboratory, will lose its vitreousness.

In summary, the wheat endosperm varies both in texture (hardness) and appearance (vitreousness). In general, high-protein hard wheats tend to be vitreous, and low-protein soft wheats tend to be opaque. However, the causes of hardness and vitreousness are different, and the two do not always go together. Hardness is caused by the genetically controlled strength of the bond between protein and starch in the endosperm. Vitreousness, on the other hand, results from lack of air

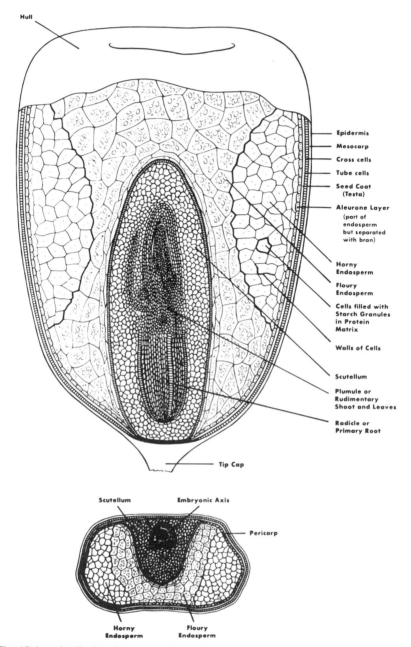

Fig. 12. Longitudinal and cross sections of a corn (maize) kernel.

spaces in the kernel. The controlling mechanism is not clear but appears to be related to the amount of protein in the sample. For example, high-protein soft wheats are more vitreous than low-protein soft wheats, and low-protein hard wheats have more *yellow-berry* (opacity) than their high-protein counterparts.

Corn

Many types of corn (or more accurately, maize) are grown around the world. This discussion is limited to the dent type; popcorn is discussed in the chapter on snack foods. Dent corn has a large, flattened seed. It is by far the largest of the common cereal seeds, weighing an average of 350 mg. The kernel (Fig. 12) is made up of four principal parts: hull or bran (pericarp and seed coat), germ, endosperm, and tip cap. For corn, "hull" is a misnomer; it is not synonymous with the hull of barley or oats but more akin to the "bran" of wheat milling terminology. However, the term "hull" is strongly ingrained in the wet-milling industry and thus will persist. The tip cap, the attachment point of the cob, may or may not stay with the kernel during shelling (removal from the cob). The botanical parts of the corn caryopsis are similar to those found in wheat. The corn kernel is quite variable in color. It may be solid or variegated and ranges from white to dark brown or purple. White and yellow are the most common colors. The hull or pericarp constitutes about 5–6% of the kernel; the germ is relatively large, constituting 10–14% of the kernel, with the remainder being endosperm.

Corn is different from wheat in that both translucent and opaque endosperm are found within a single kernel. The cellular nature of the corn endosperm is shown in Fig. 13. The cells are large with very thin cell walls.

The translucent endosperm (Fig. 14) is tightly compact, with few or no air spaces, as one might expect. The starch granules, polygonal in shape, are held together by a matrix protein. Protein bodies are quite noticeable in the photomicrograph; these have been identified as *zein* bodies. Also noticeable are indentations in the starch caused by the protein bodies. In the opaque endosperm (Fig. 15), the starch granules are spherical and are covered with matrix protein that does not contain protein bodies. The many air spaces are to be expected from its opacity. Chemical analysis of the separated opaque and translucent endosperms showed that the two contained equal amounts of protein but that the protein types varied.

In general, corn kernels are quite hard. The large number of broken starch granules in Fig. 16 shows that the bond between the protein and starch must be quite strong. The fact that water alone will not allow a good separation of protein and starch during wet-milling suggests that the bonds are different in corn and wheat. The opaque endosperm in corn is generally referred to as the "soft" endosperm.

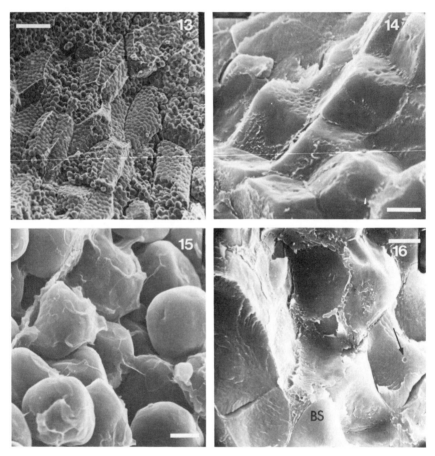

Figs. 13–16. Scanning electron micrographs of corn kernels. Each bar is 5 μm.
13. A broken kernel, showing the cellular nature of the endosperm.
14. Cross section of the vitreous part of a kernel, showing the polygonal shape of the starch granules, the indentation in the starch, and the tight compact structure.
15. Cross section of the opaque part of a kernel, showing the spherical shape of the starch granules, the protein, and the large amount of air space.
16. Cross section of the hard endosperm of a kernel, showing the starch hilum (the point from which the starch granule grew, arrow) and broken starch (BS).

While the particle size of ground corn from a mutant with a completely opaque endosperm suggests a soft endosperm and while photomicrographs of the opaque section of a normal kernel (Fig. 15) show no broken starch granules, which is compatible with a soft endosperm, it still appears prudent to call this part of the endosperm "opaque" and not assume that it is soft simply because it is opaque.

The starch granules in the opaque and translucent parts of the endosperm differ in shape. One possible explanation of why a single kernel of grain should have two starch shapes is that during the natural

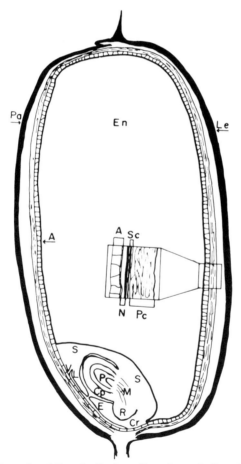

Fig. 17. A rice kernel, midlongitudinal section through the rice caryopsis: En, endosperm; Pa, palea; Le, lemma; Pc, pericarp; Sc, seed coat; N, nucellus; A, aleurone. In the germ: S, scutellum; V, ventral scale; L, lateral scale; Cp, coleoptile; P, plumule; M, mesocotyl; R, radical; E, epiblast; C, coleorhiza.

drying process the protein loses water and shrinks. The adhesion between the protein and starch is strong enough to pull the starch granules closer and closer together. At this stage, the starch granules are pliable and, as they are tightly packed, they become polygonal in shape. Further evidence of their plasticity before maturity are the indentations that the zein bodies make on the starch granules in the translucent endosperm.

In the opaque endosperm, protein distribution and amino acid composition are quite different. During drying, protein-protein bonds

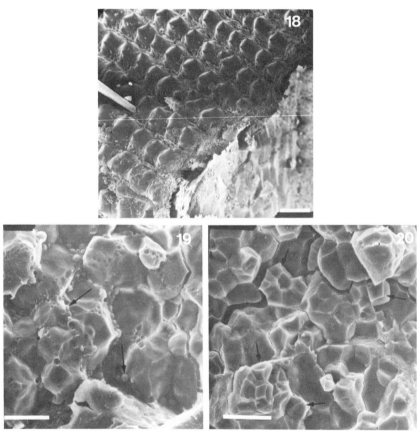

Figs. 18–20. Scanning electron micrographs of cross sections of a rice kernel.
  18. The outer surface of the rice hull. Bar is 100 μm.
  19. Compound starch granules and protein bodies (arrows) near the aleurone layer. Bar is 10 μm.
  20. Compound starch granules near the center of the kernel, with certain granules broken, showing the individual granules (arrows). Bar is 10 μm.

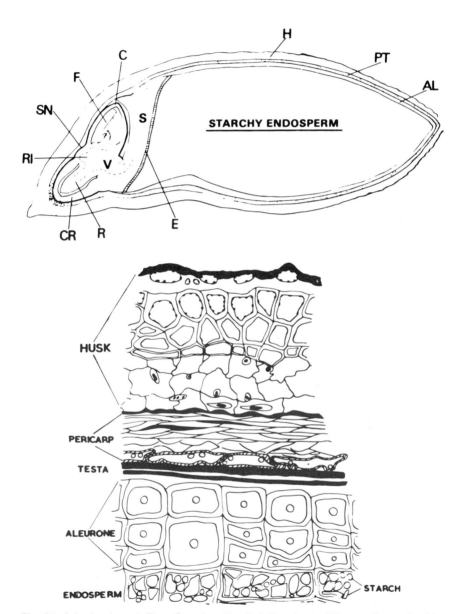

Fig. 21. A barley kernel. Top, C, coleoptile; F, foliar shoot; SN, scutellar node; RI, root initial; CR, coleorhiza; R, root; V, vascular system; S, scutellum; E, scutellar epithelial cell; H, husk; PT, pericarp-testa; AL, aleurone layer. Bottom, outer layer of the kernel.

rupture, giving intergranular air spaces and maintaining spherical starch granules. If corn is harvested before it dries, essentially all of its starch granules are spherical.

Rice

The caryopsis of rice is harvested with the hull or husk attached (Fig. 17). This is called *paddy* or *rough rice*. The hull (Fig. 18), which constitutes about 20% of the weight of rough rice, is made up of the floral envelopes, the *lemma* and *palea*. The hulls are high in cellulose (25%), lignin (30%), pentosans (15%), and ash (21%). This ash is unique in containing about 95% silica. The high amounts of lignin and silica make the rice hull of rather low value both nutritionally and commercially.

Brown rice (rice after the hull is removed) has the same gross structure as that of the other cereals. However, the caryopsis does not have a crease. It varies from 5 to 8 mm in length, and it weighs about 25 mg. Brown rice consists of a pericarp (about 2%), seed coat and aleurone (about 5%), germ (2–3%), and endosperm (89–94%). As with the other cereals, the aleurone is the outermost layer of the endosperm but is removed with the pericarp and seed coat during milling.

In general, the endosperm of rice is both hard and vitreous. However, opaque cultivars are known and some cultivars have opaque areas

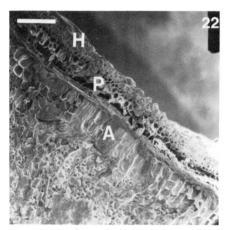
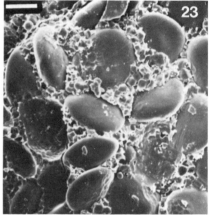

Figs. 22 and 23. Scanning electron micrographs of cross sections of a barley kernel.
22. H, hull; P, pericarp; A, multilayered aleurone cells. Bar is 100 μm.
23. Contents of an endosperm cell. Bar is 10 μm.

(called *white belly*); these are similar to yellow-berry in wheat. The opacity is caused by air spaces in the endosperm. The thin-walled endosperm cells are tightly packed, with polygonal compound starch granules and protein bodies (Fig. 19). The protein bodies are more numerous in the cells just inside the aleurone than in cells near the center of the endosperm (Fig. 20). The polygonal starch granules may be formed by compression of the starch granules during grain development. Rice and oats are the only two cereals with *compound starch granules* (i.e., a large granule made up of many small granules). The individual rice starch granules are small (2–4 μm).

Barley

Barley, like rice and oats, is harvested with the hull (or husk) intact. The tightly adhering hull consists of the lemma and palea. The caryopsis is composed of pericarp, seed coat, germ, and endosperm (Fig. 21). The aleurone cells in barley make up the outer two to three layers of cells (Fig. 22). Medium-sized kernels weigh approximately 35 mg. The aleurone of some cultivars is blue, whereas in others it is white. The endosperm cells are packed with starch embedded in a protein matrix (Fig. 23). Like wheat starch, barley starch has both large lenticular granules and small spherical granules.

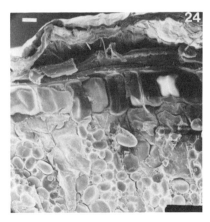
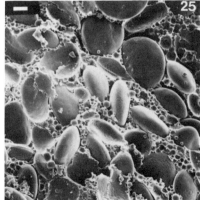

Figs. 24 and 25. Scanning electron micrographs of cross sections of a rye kernel.
24. Outer part of the kernel. Bar is 20 μm.
25. Contents of an endosperm cell. Bar is 10 μm.

Rye

The rye kernel is a caryopsis 6-8 mm in length and 2-3 mm in width. The kernel threshes free of glumes, has no hull, and is creased like wheat. Its color is grayish yellow. Like the other cereals, rye has a caryopsis consisting of pericarp, seed coat, nucellar epidermis, germ, and endosperm; the endosperm is surrounded by a single layer of aleurone cells. Scanning electron micrographs of the outer area of the grain (Fig. 24) and of the endosperm (Fig. 25) show that they are similar to those of wheat. The starch in the endosperm cells is embedded in a protein matrix. The starch, like wheat and barley starches, has large lenticular and small spherical granules.

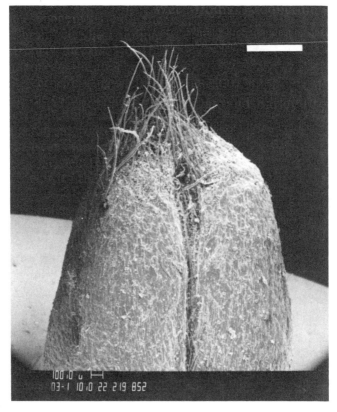

Fig. 26. Scanning electron micrograph of the outer surface of an oat groat. Bar is 500 μm.

Oats

Oats, like barley and rice, are harvested with the caryopsis enclosed in a floral envelope. The caryopsis itself is called a *groat*. The oat groat is similar in appearance to kernels of wheat or rye except that it is covered with numerous *trichomes* (hairlike protuberances, Fig. 26). The germ extends about one third of the length of the groat, being larger and narrower than the germ of wheat. The oat groat consists of pericarp, seed coat, hyaline layer, germ, and endosperm. The aleurone makes up the outer layer of the endosperm. The starchy endosperm of oats contains more protein and oil than do those of the other cereals. The starch is found as large compound granules (Fig. 27) made up of many small individual granules. The small granules are polyhedral in shape and range in size from 3 to 10 μm (Fig. 28).

Sorghum

The kernels of sorghum thresh free of hulls or glumes. They are generally spherical, range in weight from 20 to 30 mg, and may be white, red, yellow, or brown. Figure 29 shows the various parts of the sorghum kernel. Hand-dissected kernels were found to be 7.9% pericarp (presumably pericarp plus seed coat), 9.8% germ, and 82.3% endosperm. Scanning electron micrographs of the outer layers of a sorghum kernel reveal a thick pericarp, in most varieties, consisting of three layers: the *epicarp*, the *mesocarp*, and the *endocarp* (Fig. 30).

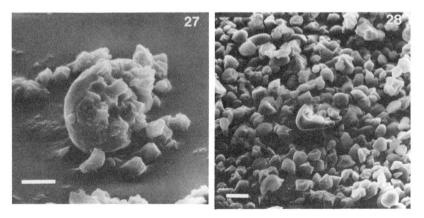

Figs. 27 and 28. Scanning electron micrographs of oat starch. Each bar is 10 μm.
27. A compound granule.
28. Isolated starch. Compound granules are destroyed during the isolation.

Unlike other cereals, many sorghum varieties contain starch granules in the pericarp. These granules, ranging in size from 1 to 4 µm, are located in the mesocarp. The endocarp layer is composed of cross and tube cells.

The mature sorghum caryopsis may (Fig. 31) or may not (Fig. 32) have a pigmented *inner integument*. The inner integument is often erroneously called a *testa* layer. The testa is the seed coat, which is joined to the outer edge of the inner integument. All mature sorghum seeds have a testa (seed coat); however, certain cultivars lack a

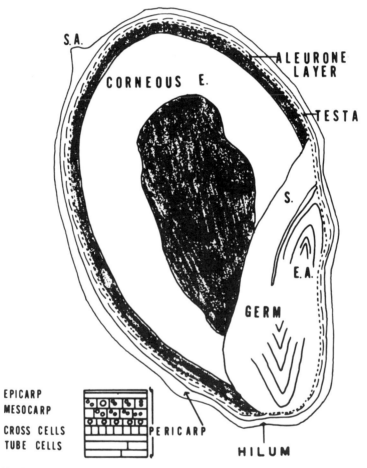

Fig. 29. A sorghum kernel. S.A., stylar area; E., endosperm; S., scutellum; E.A., embryonic axis.

pigmented inner integument. The pigmented inner integument often contains high levels of *condensed tannins*. Cultivars that have a large pigmented inner integument are called "bird-resistant" sorghums. Birds do not like the bitter tannins; however, if no other feed is available, they will eat the bird-resistant types.

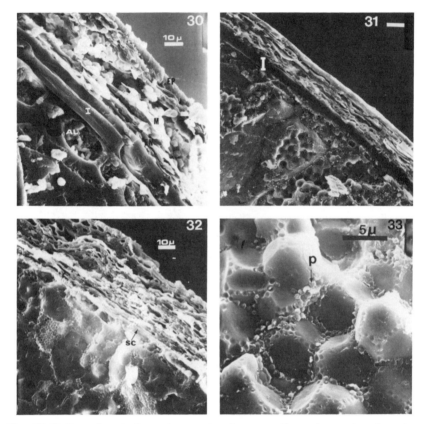

Figs. 30–33. Scanning electron micrographs of cross sections of a sorghum kernel.
30. The outer edge of the kernel, showing the epicarp (EP), mesocarp (M), endocarp (EN), inner integument (I), and aleurone cells (AL). Note the small starch granules in the mesocarp.
31. The outer edge, showing the presence of a thick pigmented inner integument (I). Bar is 20 μm.
32. A sorghum kernel containing no inner integument. The seed coat (SC) or testa is shown.
33. The vitreous part of the kernel, showing the contents of an endosperm cell. Note the lack of air spaces, the polygonal starch granules, and the protein bodies (P).

As in other cereals, the aleurone cells are the outer layer of the endosperm. In the starchy endosperm, cells containing high concentrations of protein and few starch granules are found just beneath the aleurone layer. Much of the protein is in the form of protein bodies 2–3 μm in diameter (Fig. 33). Sorghum kernels, like corn kernels, contain both translucent and opaque endosperm. The opaque endosperm has large intergranular air spaces (Fig. 34), which are responsible for its opaque appearance.

The terms "hard" and "soft" have been used to designate the vitreous and opaque areas of sorghum endosperm as well as the general appearance of kernels. However, as discussed previously for wheat, the factors determining vitreousness and physical hardness are different. Therefore, some kernels may appear vitreous but be classified soft by objective measurements. Visual determination of hardness or softness in sorghum kernels is based on the assumption that hardness and vitreousness are the same. This appears to be an unwarranted assumption.

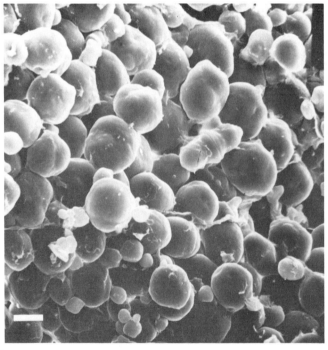

Fig. 34. Scanning electron micrograph of a cross section of the opaque part of a sorghum kernel. Note the air spaces and more or less spherical starch granules. Bar is 10 μm.

Pearl Millet

Pearl millet consists of small (average about 8.9 mg), tear-shaped kernels that are threshed clean of their hulls. They vary in color, with slate gray being most common, although yellow, white, and brown varieties are also known. The caryopsis is similar to those of the other cereals. Millet pericarp does not contain starch, as the pericarp of sorghum does, nor does pearl millet contain a pigmented inner integument. The germ in pearl millet is large (17%) in proportion to the rest of the kernel (Fig. 35). Its endosperm has both translucent and opaque endosperm, as do those of sorghum and corn. The opaque endosperm contains many air spaces and spherical starch granules (Fig. 36). The translucent endosperm (Fig. 37) is void of air spaces and contains polygonal starch granules embedded in a protein matrix. The matrix also contains protein bodies ranging in size from 0.3 to 4 μm. The protein bodies have a well-defined internal structure, as shown by the pattern of electron scattering in transmission electron micrographs.

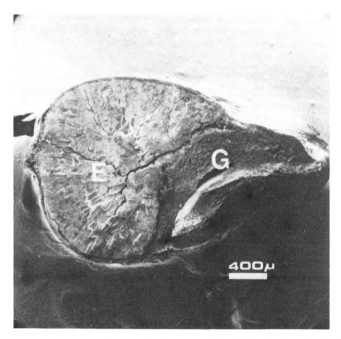

Fig. 35. Scanning electron micrograph (low magnification) of a longitudinal section of a pearl millet kernel. G, germ; E, endosperm.

 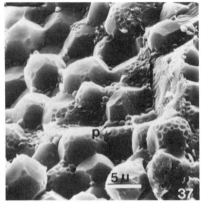

Figs. 36 and 37. Scanning electron micrographs of cross sections of a pearl millet kernel.
36. The opaque part. Note the air spaces and the spherical starch granules.
37. The vitreous part. Note the lack of air space, the polygonal starch granules, and the protein bodies (P).

Triticale

Triticale is a new cereal produced by crossing wheat (*Triticum*) and rye (*Secale*). In general morphology, the grain closely resembles its parent species. The caryopsis threshes free of glumes, is generally larger than wheat (10–12 mm in length and 3 mm in width), and weighs about 40 mg. It consists of a germ attached to an endosperm with aleurone as the outer layer; outside the aleurone are a seed coat, a pericarp, and the remains of the nucellar epidermis. Thus, triticale closely resembles the other cereal grains in structure.

The kernel has a crease that extends its full length. The yellowish brown grain is characterized by folds of the outer pericarp, apparently caused by shriveling of the grain.

Grain shriveling is a major problem with triticale. It leads to low test weight, poor appearance, and unsatisfactory milling performance. The aleurone layer in triticale is more irregular in shape than is that in wheat. The cells vary in size, and the cell wall tends to vary in thickness. In grain that is shriveled, the aleurone cells are badly distorted, and lesions have been noted in which complete sections of aleurone and associated endosperm cells are missing.

REVIEW QUESTIONS

1. What is a caryopsis?

2. Explain why some kernels or parts of kernels appear opaque or floury when a cut surface is viewed, whereas others appear glassy or vitreous.
3. Explain what is responsible for the differences in hardness of different cereals.
4. How are opacity and vitreousness related to hardness in cereals?
5. What cereal has a multiple layer of aleurone cells?
6. What are trichomes, and where are they found?
7. Which cereal has starch granules in the pericarp?
8. What cereal sometimes contains condensed tannins?
9. What percentage of the pearl millet kernel is germ?
10. Triticale was produced from the cross of what two cereals?
11. Pearl millet can vary in color; what are the common colors?
12. What cereals are usually harvested with their hulls intact?
13. About what does an average wheat kernel weigh?
14. Why does corn endosperm contain both polygonal and spherical starch granules?
15. What is paddy?

SUGGESTED READING

BECHTEL, D. B., and POMERANZ, Y. 1980. The rice kernel. Pages 73-113 in: Advances in Cereal Science and Technology, Vol. 3. Y. Pomeranz, ed. Am. Assoc. Cereal Chem., St. Paul, MN.

BRADBURY, D., CULL, I. M., and MacMASTERS, M. M. 1956. Structure of the mature wheat kernel. I. Gross anatomy and relationships of parts. Cereal Chem. 33:329-342.

BRADBURY, D., MacMASTERS, M. M., and CULL, I. M. 1956. Structure of the mature wheat kernel. III. Microscopic structure of the endosperm of hard red winter wheat. Cereal Chem. 33:361-373.

BRADBURY, D., MacMASTERS, M. M., and CULL, I. M. 1956. Structure of the mature wheat kernel. IV. Microscopic structure of the germ of hard red winter wheat. Cereal Chem. 33:373-391.

BUSHUK, W., and LARTER, E. N. 1980. Triticale: Production, Chemistry and Technology. Pages 115-157 in: Advances in Cereal Science and Technology, Vol. 3. Y. Pomeranz, ed. Am. Assoc. Cereal Chem., St. Paul, MN.

ESAU, K. 1977. Anatomy of Seed Plants, 2nd ed. John Wiley and Sons, New York.

EVERS, A. D., and BECHTEL, D. B. 1988. Microscopic structure of the wheat grain. Pages 47-95 in: Wheat: Chemistry and Technology, 3rd ed. Vol. 1. Y. Pomeranz, ed. Am. Assoc. Cereal Chem., St. Paul, MN.

GREENWELL, P., and SCHOFIELD, J. D. 1986. A starch granule protein associated with endosperm softness in wheat. Cereal Chem. 63:379-380.

HOSENEY, R. C., DAVIS, A. B., and HARBERS, L. H. 1974. Pericarp and endosperm structure of sorghum grain shown by scanning electron microscopy. Cereal Chem. 51:552-558.

HOSENEY, R. C., and SEIB, P. A. 1973. Structural differences in hard and soft wheats. Bakers Dig. 47(6):26-28, 56.

HOSENEY, R. C., VARRIANO-MARSTON, E., and DENDY, D. A. V. 1981. Sorghum and millets. Pages 71-144 in: Advances in Cereal Science and Technology, Vol. 4. Y. Pomeranz, ed. Am. Assoc. Cereal Chem., St. Paul, MN.

JULIANO, B. O., and BECHTEL, D. B. 1985. The rice grain and its gross composition. Pages 17-57 in: Rice: Chemistry and Technology, 2nd ed. B. O. Juliano, ed. Am. Assoc. Cereal Chem., St. Paul, MN.

MALOUF, R. B., LIN, W. D., and HOSENEY, R. C. 1992. Wheat hardness. II. Effect of starch granule protein on endosperm tensile strength. Cereal Chem. 69:169-173.

MARSHALL, H. G., and POMERANZ, Y. 1982. Buckwheat: Description, breeding, production, and utilization. Pages 157-210 in: Advances in Cereal Science and Technology, Vol. 5. Y. Pomeranz, ed. Am. Assoc. Cereal Chem., St. Paul, MN.

MATZ, S. A. 1969. Cereal Science. Avi Publishing Co., Westport, CT.

POMERANZ, Y., and SACHS, I. B. 1972. Determining the structure of the barley kernel by scanning electron microscopy. Cereal Chem. 49:1-4.

POMERANZ, Y., and SACHS, I. B. 1972. Scanning electron microscopy of the oat kernel. Cereal Chem. 49:20-22.

ROBUTTI, J. L., HOSENEY, R. C., and WASSOM, C. E. 1974. Modified *opaque-2* corn endosperms. II. Structure viewed with a scanning electron microscope. Cereal Chem. 51:173-180.

ROONEY, L. W., and SULLINS, R. D. 1977. The structure of sorghum and its relation to processing and nutritional value. Pages 91-109 in: Proceedings of a Symposium on Sorghum and Millets for Human Food. D. A. V. Dendy, ed. Tropical Products Institute, London.

SIMMONDS, D. H., and CAMPBELL, W. P. 1976. Morphology and chemistry of the rye grain. Pages 63-110 in: Rye: Production, Chemistry, and Technology. W. Bushuk, ed. Am. Assoc. Cereal Chem., St. Paul, MN.

Starch

CHAPTER 2

Cereal grains store energy in the form of starch. The amount of starch contained in a cereal grain varies but is generally between 60 and 75% of the weight of the grain. Thus, much of the food that humans consume is in the form of starch, an excellent source of energy.

In addition to its nutritive value, starch is important because of its effect upon the physical properties of many of our foods. For example, the gelling of puddings, the thickening of gravies, and the setting of cakes are all strongly influenced by properties of starch. Starch is also an important industrial commodity, particularly in the papermaking industry.

The Starch Granule

Starch is found in plants in the form of granules. In cereals and other higher plants, granules are formed in *plastids*. Those plastids that form starch are called *amyloplasts*. In the cereals with simple starch granules (wheat, corn, rye, barley, sorghum, and millets), each plastid (amyloplast) contains one granule. In rice and oats, which contain compound starch granules, many granules are found in each amyloplast.

Wheat, rye, and barley have two types of starch granules, the large, lenticular (lens-shaped) and the small, spherical. In barley, the lenticular granules are formed during the first 15 days after pollination. The small granules, representing about 88% of the total number of granules, appear 18–30 days after pollination. Plastids in barley and wheat each initially form a large lenticular starch granule. Later, the plastids form evaginations in which small granules are formed. These much smaller amyloplasts are separated from the mother plastid by constriction.

In most plants, sucrose is the most common sugar. In developing cereal endosperm tissue, a high positive correlation exists between sucrose concentration and rate of starch synthesis. From an osmotic standpoint, it is a great advantage for the plant to store its excess energy as starch, which is insoluble and of extremely large molecular weight, rather than as sucrose. Because starch is synthesized in plastids, those structures must possess all the enzymes necessary for granule formation. The starch grows by apposition. The new layer deposited on the outside of the granule varies in thickness, depending upon the amount of carbohydrate available at the time. These layers become apparent after treatment of the starch with dilute acid or enzymes (Fig. 1). In potato starch, the layers are obvious and easily seen in the intact starch under a light microscope (Fig. 2). The nature and cause of the layers in starch is still uncertain, but they presumably represent growth rings.

CRYSTALLINITY

The fact that starch is a semicrystalline material has been clear since the classic work of Katz and his collaborators in the 1930s. Intact

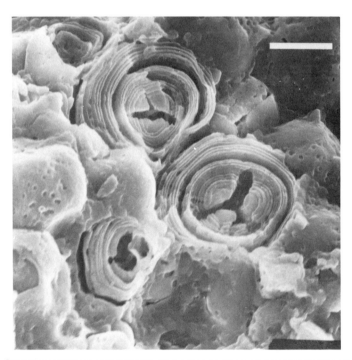

Fig. 1. Scanning electron micrograph of a cross section of a sorghum kernel treated with α-amylase. Note the rings in the broken starch granules. Bar is 10 μm.

starch granules give three types of X-ray patterns (designated A, B, and C). Most cereal starches give the A pattern (Fig. 3); potato, other root starches, and retrograded starch give a B pattern (Fig. 3); and smooth pea and bean starches give the C pattern, which is an intermediate form, probably due to mixtures of the A and B types. The B pattern of potato starch can be converted to an A pattern by heat-moisture treatments. Small dextrins (12–15 glucose units) could yield any of the three patterns, depending on their crystallization conditions. X-ray diffraction is a powerful tool for starch investigations.

X-ray diffraction of crystalline amylose helical inclusion compounds (discussed later) results in the V amylose pattern (Fig. 3). This pattern is not found in naturally occurring starch but only after gelatinization and complexing with lipid or related compounds.

BIREFRINGENCE

When viewed in polarized light, starch granules show *birefringence* in the form of the typical "maltese cross" (Fig. 4). The property of birefringence is brought about because the starch granule has a high degree of molecular order. This must not be confused with crystal-

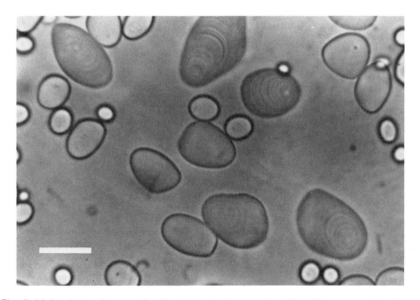

Fig. 2. Light photomicrograph of potato starch granules. Note the concentric rings. Bar is 10 μm.

linity—things can be very ordered and yet not be crystalline. If they are highly ordered, they will be birefringent. For example, the cellulose in a piece of paper is semicrystalline and the crystallites themselves are birefringent (ordered). However, because the crystallites are randomly oriented, the paper as a whole is not birefringent.

Chemical Composition

Starch is composed essentially of glucose. Although it may contain a number of minor constituents, these occur at such low levels that it can be debated whether they are trace constituents of the starch or contaminants not completely removed during the isolation process. Even though these minor constituents are found in small amounts in the starch, they can and do affect the starch properties.

Cereal starches contain low levels of fats. The lipids associated with starch are generally polar lipids, which require polar solvents such as methanol-water for their extraction. Generally, the level of lipids in cereal starch is between 0.5 and 1%. Noncereal starches contain essentially no lipids.

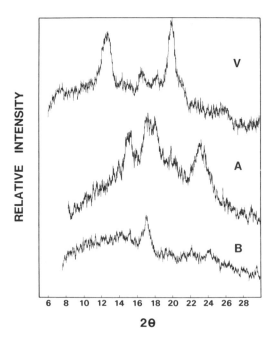

Fig. 3. X-ray diffraction patterns of starch, showing the A, B, and V patterns.

Besides low levels of other minerals, starches contain phosphorus and nitrogen. In the cereals, most of the phosphorus is in the form of phospholipids. Phosphorus in potato starch is known to be esterified to glucose residues, but this is apparently not true with the cereal starches. All starches also contain low levels of nitrogen ($<0.05\%$); part of this is from the lipids, and the remainder may be proteinaceous, perhaps remnants of enzymes involved in starch synthesis.

Starch is basically polymers of α-D-glucose. Chemically, at least two types of polymers are distinguishable: amylose, an essentially linear polymer, and amylopectin, which is highly branched.

AMYLOSE

Amylose is generally assumed to be a linear polymer of α-D-glucose linked α-1,4. The molecular weight of amylose is around 250,000 (1,500 anhydroglucose units) but varies quite widely, not only between species of plants but also within a species, and depends upon the plant's stage of maturity. Although the polymer is generally assumed to be linear, this appears to be true for only part of the amylose; the remainder appears to have a low degree of branching.

When amylose is leached from starch by heating slightly above the starch's *gelatinization temperature*, the amylose solubilized is essentially

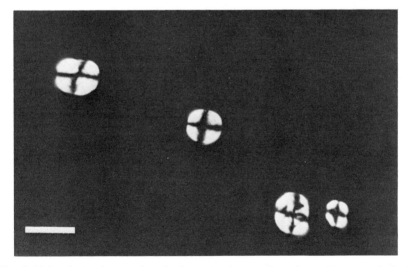

Fig. 4. Light photomicrographs of wheat starch granules taken with crossed nicol prisms and showing the maltese cross. Bar is 10 μm.

linear. As the leaching temperature is increased, amylose of higher molecular weight and more branching is extracted. Both enzymatic studies and studies of viscosity have indicated that the branches are of long-chain length, with the side chains containing hundreds of glucose residues (Fig. 5). The branch points are α-1,6 bonds, the same as those found in amylopectin. The branches on amylose are so long and so few that in many ways the molecule acts as an unbranched entity.

The long, linear nature of amylose gives it some unique properties, for example, its ability to form complexes with iodine, organic alcohols, or acids (Fig. 6). Such complexes are called *clathrates* or *helical inclusion compounds*. Amylose can be precipitated from a solution of starch (solubilized with potassium hydroxide or dimethyl sulfoxide) by the addition of *n*-butyl alcohol. The alcohol and the amylose form an insoluble complex, the nature of which is similar to that formed by iodine with amylose. The well-known blue color given by iodine and starch is thought to be due to polyiodide ions in the central core of the amylose helix.

The long linear nature of amylose is also responsible for its tendency to associate with itself and precipitate from solution. The amylose will

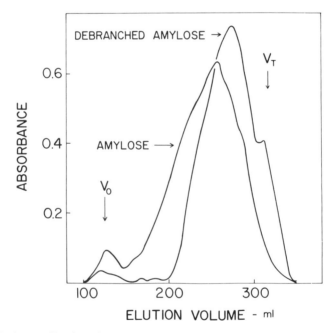

Fig. 5. Elution profile of amylose and debranched amylose from Sepharose CL-2B.

readily crystallize from solution or *retrograde*. Retrogradation is a term used to denote crystallization, including crystallization in starch gels. Because of its strong tendency to associate, amylose is difficult to work with. It will stay in solution if the pH is kept high (for instance, in $1N$ KOH), as small positive charges are induced on the OH groups and these charges on adjacent chains repel each other.

AMYLOPECTIN

Like amylose, amylopectin is composed of α-D-glucose linked primarily by α-1,4 bonds. Amylopectin is branched to a much greater extent than

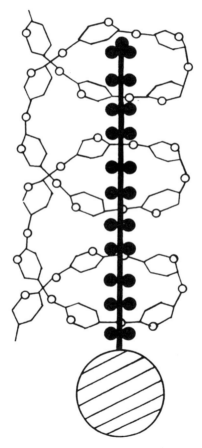

Fig. 6. Depiction of an organic acid forming a clathrate with amylose.

is amylose, with 4–5% of the glycosidic bonds being α-1,6 bonds. This level of branching means that, on the average, the unit chain in amylopectin is only 20–25 glucose units long. The molecular weight of amylopectin has been reported to be as high as 10^8. It is truly a huge molecule, one of the largest found in nature. If the above figure for molecular weight is correct, then the molecule has 595,238 glucose residues ($10^8/168$; 168 is the molecular weight of an anhydroglucose unit) or 29,762 chains with an average degree of polymerization of 20.

Amylopectin is thought to be randomly branched. The molecule has three types of chains (Fig. 7): A-chains, composed of glucose linked α-1,4; B-chains, composed of glucose linked α-1,4 and α-1,6; and C-chains, made up of glucose with α-1,4 and α-1,6 linkages plus a reducing group. Thus, A-chains do not carry branches and B-chains do; the C-chain is branched and also has the only reducing group (unbound C-1) in the molecule.

The structure of amylopectin is best examined by using a series of enzymes that partially degrade the molecule in very specific ways. One such enzyme is β-amylase, which attacks at the nonreducing end of starch chains and breaks every second α-1,4 bond. β-Amylase cannot pass a branch point on the starch chain. It thus reduces a linear chain to maltose (two glucose molecules linked α-1,4) but leaves two or three glucose residues at the branch point, depending upon whether the original chain had an even or odd number of glucose residues outside

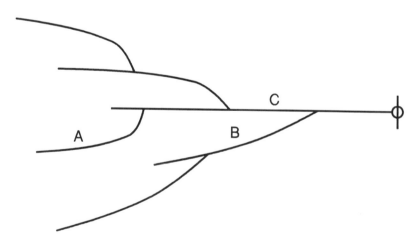

Fig. 7. Identification of A, B, and C chains of amylopectin. The lines represent glucose residues linked α-1,4 and α-1,6 at the branch points. ϕ, reducing group.

the branch point. Amylopectin is degraded about 55% by β-amylase. The products are maltose and the large residue, *β-limit dextrin*. The β-limit dextrin still has a very large molecular weight (10^4).

Two other enzymes that have been helpful in determining the structure of amylopectin are the debranching enzymes pullulanase and isoamylase. Both hydrolyze the α-1,6 bonds but not the α-1,4 bonds. Thus, treating amylopectin with either of these enzymes breaks all the α-1,6 bonds (branches) and leaves relatively short linear chains. By determining the reducing power per unit weight, the *average chain length* can be calculated. Amylopectin has a average chain length of about 25. The distribution of the chains can be seen by subjecting them to gel filtration on Bio-Gel P-10 or a similar material. The distribution is bimodal, with most (by weight) of the chains being about 19 glucose units in length and the others about 60 glucose units.

An important difference between the debranching enzymes is that pullulanase debranches side chains of two or more glucose units, whereas isoamylase requires at least three glucose units. This difference in the ratio specificity allows us to calculate an *A-chain to B-chain ratio*. A β-limit dextrin is prepared and treated with isoamylase, which cleaves all the B-chains and half the A-chains (those having an odd number of glucose units and thus reduced to maltotriose by β-amylase). The same β-limit dextrin is then treated with pullulanase, which cleaves all the A- and B-chains. The number of chains can be determined by the reducing power generated by the cleavage. Subtracting the values obtained with isoamylase from those obtained with pullulanase gives a value equal to one half of the A-chains. Multiplying by two gives the A-chains, and subtracting the A-chains from the A- plus B-chains (the pullulanase digest) gives the B-chains. Of course, dividing the two gives an A- to B-chain ratio. However, each mathematical step increases the error. Thus, with the number of steps required, the values are meaningless unless the reducing power can be obtained with extreme precision. An alternative and much better method to determine the A-B ratio is to produce the β-limit dextrin, debranch it with pullulanase, and subject it to gel filtration chromatography to separate the component polymers. The sum of the maltose molecules plus the maltotriose molecules gives the number of A-chains. The number of larger molecules gives the B-chains. The ratio of the reducing power between the large chains (larger than maltotriose) and small chains (maltotriose and maltose) gives the A-B ratio (each reducing group comes from one chain). The A-B chain ratio is one method of characterizing amylopectin.

FRACTIONATION OF STARCH

Two general techniques are used to fractionate starch into its components of amylose and amylopectin. The amylose can be selectively leached from granules that have been heated to slightly above the gelatinization temperature. An increased leaching temperature solubilizes amylopectin in addition to amylose, and thus further purification is required. The fractionations (separations) achieved by leaching are not quantitative. Pretreatment of starch with hot aqueous butanol before aqueous leaching makes amylopectin less soluble and gives a higher yield of amylose.

The second method is to completely disperse the granule and then separate the components. Cereal starches are particularly difficult to disperse completely; several hours at autoclave temperatures (130°C) are required. Care must be taken that the starches are not degraded under those conditions. For example, the starch should be defatted, buffered, and protected from oxygen.

A number of pretreatments, for example, use of liquid ammonia, dimethyl sulfoxide, or dilute base, have been used to help completely disperse the starch. After the starch is completely dispersed, the most common separation technique is to precipitate the amylose as its *n*-butanol or thymol complex. Several reprecipitations are necessary to obtain pure amylose. The amylopectin can be recovered by lyophilization or precipitation with alcohol.

Organization of the Starch Granule

How the amylose and amylopectin molecules are arranged in the starch granule is not known with great certainty. Recent work provides sufficient information, however, to reach some tentative conclusions. It appears that most, if not all, of the molecules in a starch granule are oriented at a right angle to the granule's surface. As mentioned earlier, the starch granule is partially crystalline. The amylopectin molecules in the granule appear to be the ones that are partially crystalline. Waxy starches (100% amylopectin) have the same degree of crystallinity and X-ray type as do regular starches containing both amylose and amylopectin. The model proposed by French and Kainuma (Fig. 8) shows the direction of growth of the amylopectin molecule but accounts only for the amylopectin. The location of amylose in the granule is not known. It also shows that the length of the crystal is limited by the length of the small chain (20 glucose units) and that the crystal grows at right angles to the length of the molecule. When studying this model, one should keep in mind that the starch granule is only about 30% crystalline.

The outer chains of amylopectin molecules appear to occur as a double helix of about 50 Å in length. Whether all or only a part of these chains occurs as a double helix is not clear.

The crystallinity of starch granules can be destroyed by mechanical means. Ball milling starch at room temperature will eventually completely destroy both the birefringence and the X-ray pattern. In general, we do not have a good explanation for this phenomenon. It is quite clear that much remains to be learned about the structure of the starch granule.

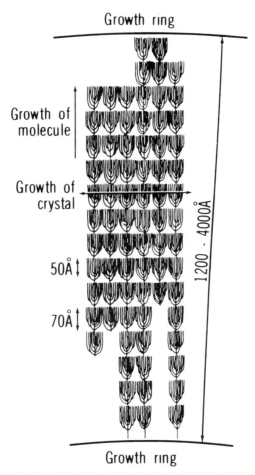

Fig. 8. Model of the structure of amylopectin in starch granules.

Starches from Different Cereals

The starches from different cereals vary widely in their size, shape, and gelatinization properties (Table I). The ratio of amylose to amylopectin is relatively constant, at about 23 ± 3% amylose. However, mutants that have greatly changed ratios are known in a number of cereals. These mutants are discussed in more detail later.

Wheat, barley, and rye all have two types and sizes of starch granules (Fig. 9): the large (25–40 μm) lenticular and small (5–10 μm) spherical granules. Scanning electron micrographs of isolated wheat starch granules are shown in Fig. 10. Most evidence suggests that the chemical composition and properties of the two types of granules are essentially the same. Of course, one obvious difference is the small granule's much larger surface area per unit mass. The literature on the rate of enzymatic attack of the two sizes of granules is confusing, with conflicting reports. The gelatinization properties of starches from these three grains are also very similar; 50% loss of birefringence occurs at about 53°C in excess water. Triticale starch appears to also be very similar to these starches as well.

The starch granules of corn and sorghum are very similar to each other in size, shape, and gelatinization properties. They average about 20 μm in diameter (Fig. 11), and their shape varies from polygonal to almost spherical (Fig. 12). Starch granules in cells near the outside of the kernel (i.e., in the vitreous endosperm) tend to be polygonal, whereas those in cells from the center of the kernel (in the opaque

TABLE I
Properties of Certain Starches[a]

Source of Starch	Gelatinization Temperature Range (°C)	Granule Shape	Granule Size (μm)
Barley	51–60	Round or elliptical	20–25 2–6
Triticale	55–62	Round	19
Wheat	58–64	Lenticular or round	20–35 2–10
Rye	57–70	Round or lenticular	28
Oats[b]	53–59	Polyhedral	3–10
Corn	62–72	Round or polyhedral	15
Waxy corn	63–72	Round	15
Sorghum	68–78	Round	25
Rice[b]	68–78	Polygonal	3–8

[a] From D. R. Lineback. 1984. Baker's Dig. 58(2):16-21. Used by permission.
[b] Individual, not compound, granules.

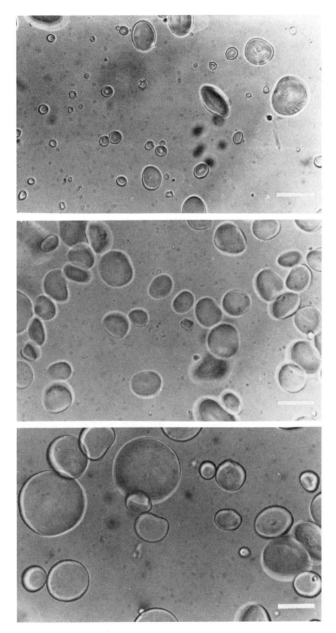

Fig. 9. Light photomicrographs of wheat (top), barley (middle), and rye (bottom) starch granules. Bar is 10 μm.

endosperm) tend to be spherical (see Figs. 14 and 15 of Chapter 1). As far as is known, the properties of the differently shaped granules are the same. Pearl millet starch is also similar to that of corn and sorghum, except that its granules are smaller, averaging about 12 μm in diameter (Fig. 13). All three starches have a 50% gelatinization temperature of about 67°C, somewhat higher than that of wheat, barley, and rye.

Individual starch granules of rice and oats are similar in that they are small (2–5 μm), are polygonal in shape, and occur in the grain as compound granules (Fig. 14). The compound granules (see Figs. 20 and 27 of Chapter 1), however, are quite different; those of oats are large and spherical and those of rice are smaller and polygonal. Also, the two starches differ quite widely in their gelatinization properties. Oat starch gelatinizes at a relative low temperature (50% at ~55°C) and rice starch at a higher temperature (50% at ~70°C).

In corn, sorghum, barley, and rice, mutants that have starches with essentially 100% amylopectin have been discovered. The starches are

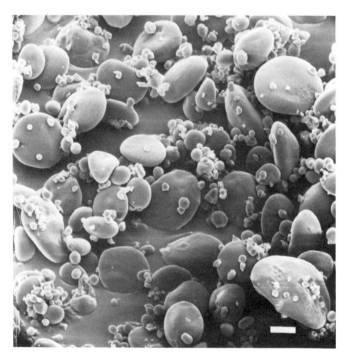

Fig. 10. Scanning electron photomicrographs of isolated wheat starch granules. Bar is 10 μm.

called *waxy* starches, and the cereals that contain them are called waxy corn, waxy barley, etc. Mutants that have starches with unusually high levels of amylose are also known. Certain lines of corn have starch containing 70% amylose. Such cereals are known as *amylotypes*. Starches

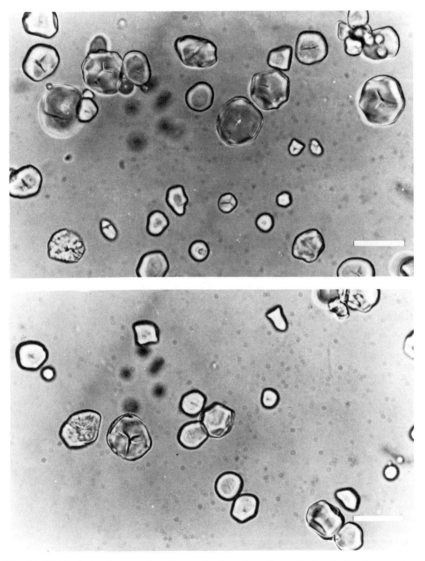

Fig. 11. Light photomicrographs of corn (top) and sorghum (bottom) starch granules. Bar is 10 μm.

varying in the ratio of amylose to amylopectin have not been found for all cereals, and wheat is a notable example of a cereal in which the ratio does not vary.

Heating Starch in Water

The changes that starch undergoes when it is heated with water are responsible for the unique character of many of our foods. Some obvious examples are the viscosity and mouth-feel of gravies and puddings and the texture of gum drops and pie filling. Just as important, but not quite as obvious, is the effect of those changes on baked products. All baked products "set," that is, reach a temperature at which the dough or batter can no longer expand under the gas pressure generated by the increasing temperature. The changes that starch undergoes are responsible, at least partially, for that setting.

Most systems used to study the interaction of starch, water, and temperature work only at very dilute starch-in-water concentrations.

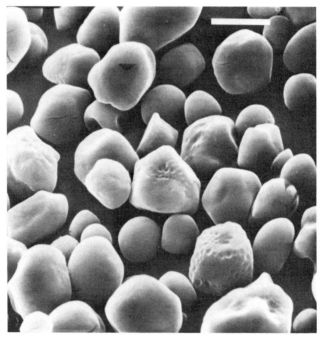

Fig. 12. Scanning electron photomicrographs of isolated corn starch granules. Bar is 10 μm.

These conditions are quite different from the concentrated starch-in-water systems found in baked products. Little or no data suggest that results obtained in dilute systems can be used to understand what is occurring in concentrated food systems. But although the changes that starch undergoes in dilute systems may or may not apply to our food systems, it is still important to understand them.

When starch is placed in water, the granule is freely penetrated by water, or for that matter by most small molecules (up to a molecular weight of about 1,000). With hydration, the starch can hold about 30% of its dry weight as moisture. The granule swells slightly; its volume increase is generally considered to be about 5%. The volume change and water absorption are reversible, and heating the system to just below its gelatinization temperature will not bring about any other changes. However, heating to higher temperatures results in irreversible changes, which can be shown with an *amylograph.*

The amylograph measures the relative viscosity of a system as it is heated at a constant rate (1.5°C/min). The amylograph is not sensitive

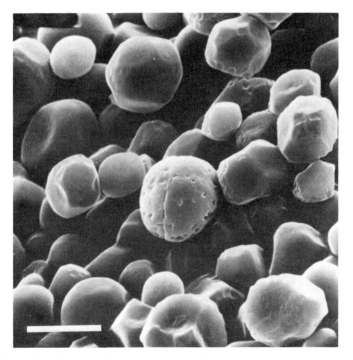

Fig. 13. Scanning electron photomicrographs of isolated pearl millet starch granules. Bar is 5 μm.

enough to measure small changes in relative viscosity; therefore, carboxymethyl cellulose is often added to the buffer (the official method calls for the use of a phosphate buffer, pH 6.8) to produce a baseline viscosity that is in the instrument's measurement range. When wheat

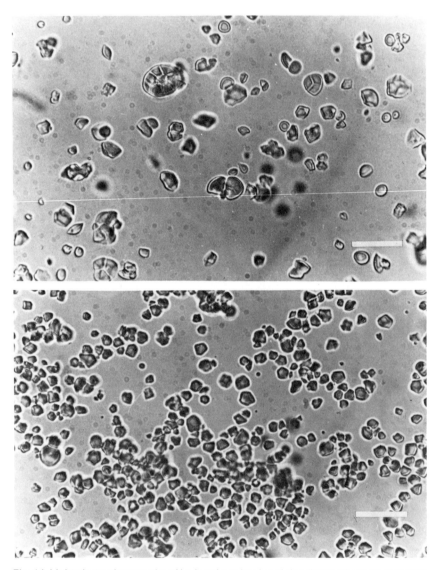

Fig. 14. Light photomicrographs of isolated oat (top) and rice (bottom) starch granules. Bar is 10 μm.

starch is added to such a buffer, changes occur as the temperature is increased (Fig. 15). Between 50 and 57°C, viscosity increases; this coincides with the loss of granular birefringence. Starch gelatinization is defined as the loss of birefringence. Continued heating in excess water gives rise to an additional, larger increase in viscosity. A new instrument that gives similar results in a much shorter time is the rapid viscoanalyzer.

The viscosity increase that occurs when starch is heated in water is the result of the starch taking up water and swelling substantially. With continued heating, the starch granule becomes distorted, and soluble starch is released into the solution. The soluble starch and the continued uptake of water by the remnants of the starch granules are responsible for the increase in viscosity. Those changes that occur after starch gelatinization (loss of birefringence) are termed *pasting*. Solubilization of the starch is continuous during this process. In excess water, the granule is not completely soluble until a temperature in excess of 120°C. Thus, in any food system we would never approach complete pasting or complete solubilization of starch.

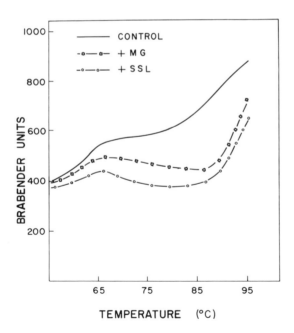

Fig. 15. Amylograph curve prepared with buffer containing carboxycellulose. MG, monoglycerides; SSL, sodium stearoyl lactylate.

In the amylograph, as in all food systems except those cooked under pressure, the temperature cannot materially exceed 100°C or the system will boil. So heating in the amylograph is discontinued at 95°C, and the temperature is held at 95°C for 1 hr. The starch is said to be *cooked*. Actually, relatively little change occurs in the nonsoluble starch. One property of starch yet to be adequately explained is that solubilization appears to be controlled mostly by temperature and not by an interaction of time and temperature. Holding starch at a specific temperature for a period of time does not increase its solubility; the temperature must be raised or the sample stirred or otherwise sheared to increase the solubility.

As shown in the stylized amylogram in Fig. 16, the viscosity of the starch system decreases markedly during the 1 hr cook at 95°C. The decrease in viscosity is caused by the molecules of soluble starch orienting themselves in the direction that the system is being stirred. This phenomenon, called *shear thinning*, is an important property of starch pastes. If you want to make a thick soup, you must not stir excessively or pump the paste through a pipe, as in both cases shear thinning will occur, giving a lower viscosity. Different starches vary in the amount of shear thinning that they show. Generally, the more soluble the starch, the more it will thin on shearing.

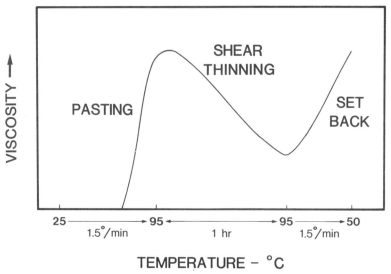

Fig. 16. Stylized amylogram for a starch-water system, showing pasting, shear thinning, and setback.

After the 1-hr heating period at 95°C, the amylograph procedure has a controlled cooling (1.5°C/min) from 95 to 50°C. This gives rise to a rapid increase in viscosity. The increase is referred to as *setback*. Starches also vary in the amount of setback. The phenomenon is caused by a decrease of energy in the system that allows more hydrogen bonding between starch chains and thus increased viscosity.

Starches from different species of cereals vary in their action when heated in water. This is illustrated in the amylograms of wheat and corn starches shown in Fig. 17.

Gelling and Retrogradation

When a starch paste is allowed to cool, it forms a gel. This statement raises two questions: What is a gel and what do we mean by starch paste? When starch granules have been heated with sufficient water and to a temperature high enough so that they gelatinize (lose

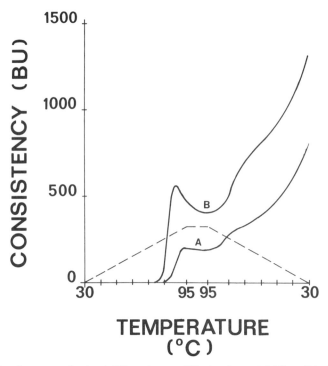

Fig. 17. Amylograms of wheat (A) and corn (B) starches at 7.5% solids in water. The dashed line is the temperature profile. BU, Brabender units.

birefringence) and part of the starch is solubilized, they are said to be pasted. Thus, a *starch paste* can vary from gelatinized granules with only a small amount of soluble starch to a system in which practically all the starch is soluble and essentially no remnants of granules can be found.

Simply stated, a *gel* is a liquid system that has the properties of a solid. Some common examples are gelatins, pie fillings, and puddings. In gels, a small amount of solid material controls a large amount of water. It is amazing that water does not leak out of the gels when they are left standing. Calculations show that the distances between the starch chains in gels are very large compared to the size of the water molecule. Studies with diffusing solutes show that the water in gels has properties essentially equal to those of pure water. However, the water is held in the gels. We do not understand the forces involved in this, but presumably hydrogen bonds are involved. The picture would be clearer if we understood the structure of water.

One can visualize the gel as starch chains with layers of water molecules attached by hydrogen bonding. As the starch paste is cooled, the starch chains become less energetic and the hydrogen bonds become stronger, giving a firmer gel. As a gel ages or if it is frozen and thawed, the starch chains have a tendency to interact strongly with each other and thereby force water out of the system. The squeezing of water out of the gel is called *syneresis*.

Longer storage gives rise to more interaction between the starch chains and eventually to formation of crystals. This process, called *retrogradation* is the crystallization of starch chains in the gel. Because the crystalline areas differ from the noncrystalline areas in their refractive index, the gel becomes more opaque as retrogradation progresses. In addition, it becomes more rigid or rubbery, perhaps partially as a result of crystallization and partially just from the interaction of the starch chains.

Heating Starch in Limited Water

It has always been, and still is, difficult to study starch in limited water systems. Yet most, if not all, of our food systems are limited in water (i.e., have less than a 2:1 ratio of water to starch). A bread dough, for example, contains about equal amounts of starch and water. The recent use of differential scanning calorimetry (DSC) has been quite beneficial for studying starch in such concentrated systems.

DSC measures heat flow as a function of temperature. When starch is heated in excess water (water-starch, 2:1), a sharp endothermic peak is obtained (Fig. 18a). The start of the peak (where it deviates from the base line) corresponds to the start of birefringence loss. The area under the curve is a measure of the energy (*enthalpy*, ΔH) required for the transition from an ordered to a disordered state (i.e., for the crystalline area to melt). The end point of the loss of birefringence and the end of the peak are not quite the same, as there is a considerable lag in the DSC. However, in general, the two correlate well.

For wheat starch in excess water, the loss of birefringence occurs over approximately a 7°C temperature range. For any one starch granule, the temperature range is much smaller, generally 1.5°C. Biological variability means that each starch granule's structure is unique. Consequently, it may start to lose birefringence at a slightly

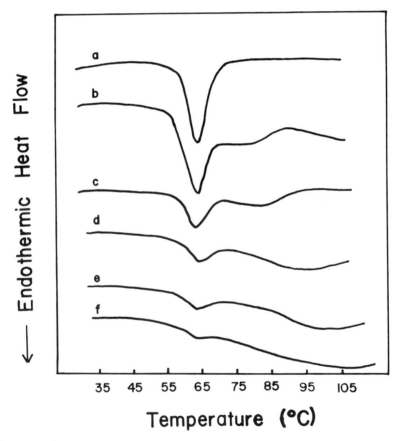

Fig. 18. Differential scanning calorimeter thermograms of wheat starch heated with different water-to-starch ratios: a, 2.0; b, 1.0; c, 0.75; d, 0.5; e, 0.44; f, 0.35.

different temperature. As the amount of water in the sample is reduced, the DSC peak widens and becomes clearly bimodal (Fig. 18). At low water contents (water-starch, 0.35:1) the loss of birefringence, as reflected by the endothermic peak in DSC, occurs over a temperature range of more than 30°C (Fig. 18f). The slow loss of birefringence can also be seen directly in light photomicrographs (Fig. 19). The amount of water available to the starch does not affect the temperature at which birefringence starts to be lost but does greatly affect the completion of the process. It is also clear that heating starch in excess water leads to swelling and distorting of the starch granules (Fig. 20). Both swelling and distortion are much less in limited water.

Modified Starches

Granular starches can be modified by chemical reaction to change their properties. The more common modifications and how they modify starch properties are discussed in this section.

ACID-MODIFIED

The oldest type of modification is treatment with acid to produce an *acid-modified starch*. The early work with acid modification was

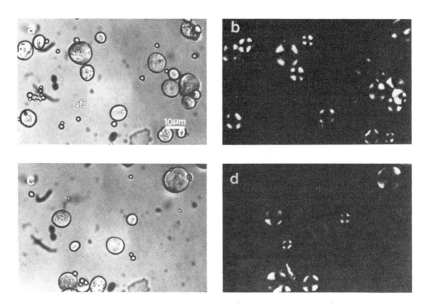

Fig. 19. Brightfield (a and c) and polarizing (b and d) micrographs of wheat starch heated in a differential scanning calorimeter to 67°C (a and b) and 87°C (c and d) at a starch-to-water ratio of 0.35.

done in the late 1800s by Lintner and Naegeli. Acid-treated starches are today referred to as *lintnerized* or Naegeli starches. The industrial preparation of today's acid-modified starches consists of treatment of a fairly concentrated starch slurry in 1–3% hydrochloric acid at about 50°C for 12–14 hr. After treatment, the slurry is neutralized and the starch recovered by filtration.

During the treatment, the acid freely penetrates the amorphous parts of the starch granule and hydrolyzes glucosidic bonds. The acid cannot penetrate the crystalline areas, perhaps because of the double helix, and they remain intact. The major effect of the acid is to reduce the molecular weight of the starch molecules while leaving the crystalline structure of the granule intact.

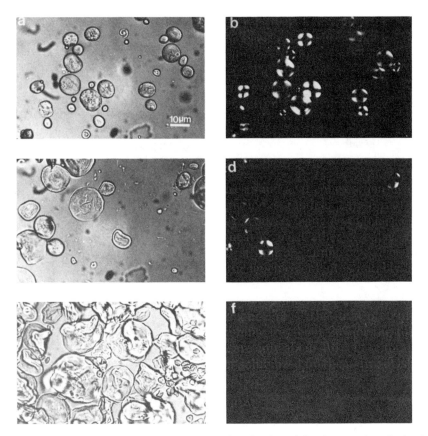

Fig. 20. Brightfield (a, c, and e) and polarizing (b, d, and f) micrographs of wheat starch heated in a differential scanning calorimeter to 57°C (a and b), 64°C (c and d), and 67°C (e and f) at a water-to-starch ratio of 2.0.

Upon heating in water, the modified granules fragment more and swell less. The gelatinization temperature range is increased, presumably because the amorphous chains cannot assist in the melting of the crystalline areas. Upon gelatinization, the starch becomes much more soluble than would untreated starch. As a result of acid treatment (hydrolysis of chains), the paste viscosity is much less than that for the native starch. Because the starch chains remaining after acid treatment are smaller (shorter in chain length), they tend to associate with each other more easily and thus form a rigid gel upon cooling. Common examples of such gels are jelly beans and other gum candies.

CROSS-LINKED

Simply stated, *cross-linking* is the covalent bonding of two starch molecules to make a larger molecule. The linking is accomplished by forming a diester with phosphoric acid ($POCl_3$ is the most common reagent) or by forming an ether bond (epichlorohydrin is the common reagent). The linkages are illustrated in Fig. 21. Although we generally consider the cross-link to occur between molecules, it could also occur within the large amylopectin molecule. However, such bonds within the molecule would not have a great effect on the starch's properties.

High levels of cross-linking increase the starch's gelatinization temperature. The more the cross-linking, the higher the temperature required for gelatinization. Highly cross-linked starches can be prepared that do not gelatinize when they are boiled in water or sterilized in an autoclave. Such modified starches are useful as dusting starches for surgeons' gloves. The starch can be sterilized, works well to protect the surgeon's hands from sticking to the gloves and, if accidently dumped into the wound, will be digested with no ill effects.

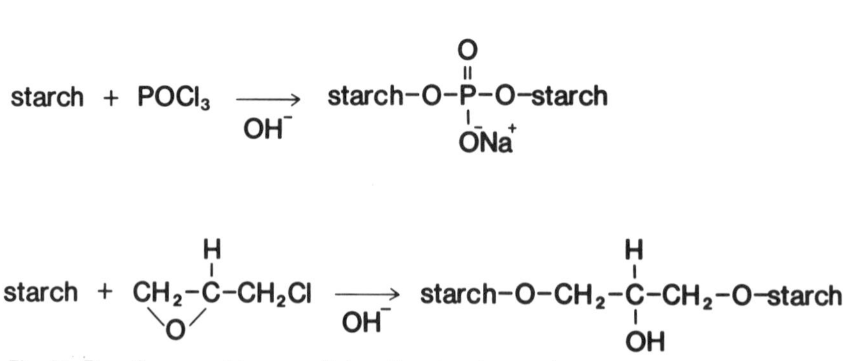

$$\text{starch} + POCl_3 \xrightarrow{OH^-} \text{starch}-O-\underset{\underset{ONa^+}{|}}{\overset{\overset{O}{\|}}{P}}-O-\text{starch}$$

$$\text{starch} + CH_2-\underset{\underset{}{}}{\overset{\overset{H}{|}}{C}}-CH_2Cl \xrightarrow{OH^-} \text{starch}-O-CH_2-\underset{\underset{OH}{|}}{\overset{\overset{H}{|}}{C}}-CH_2-O-\text{starch}$$
(with epoxide O between CH₂ and C)

Fig. 21. Reactions used to cross-link native starch granules.

Starch for use in food systems is generally cross-linked to a small extent. The amount of reaction is designated by *degrees of substitution* (DS). A DS of 1 indicates an average of one substituent for each anhydroglucose residue. Thus, the maximum DS is 3 for linear starch and slightly less for amylopectin. For food use, the DS is generally between 0.01 and 0.1.

Low levels of cross-linking do not significantly change the gelatinization temperature of the starch but do materially change its pasting properties. Cross-linked starch swells less and is less soluble than its unmodified counterpart (Fig. 22). Thus, cross-linked starch gives a lower viscosity upon pasting. One of the advantages of cross-linked starch is that it is not as subject to shear thinning as is unmodified starch. Because the starch solubilizes less, it shears less and thus gives a more viscous paste after stirring or pumping.

Another important use of cross-linked starch is to produce viscous systems in acidic media. An example is thickening a cherry pie filling. The acidity from the cherries is sufficient to speed hydrolysis of the α-1,4 glucosidic bonds in the starch during baking and thus produce a thin pie filling. Cross-linking does not stabilize the bonds to the acid; however, with sufficient cross-linking, the starch swelling is greatly

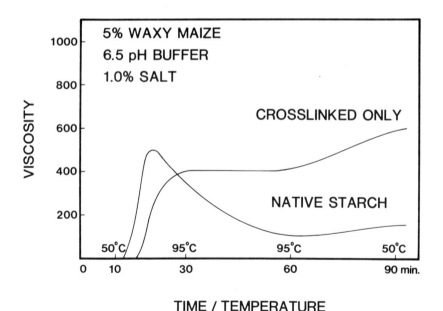

Fig. 22. Amylographs showing the effect of cross-linking on pasting properties.

restricted and, as the acid hydrolyzes the bonds, viscosity increases instead of decreasing. Therefore, if one starts with the right degree of cross-linking and obtains a constant amount of hydrolysis during baking, one can still end up with a thick pie filling. The change in viscosity as a function of pH is shown in Fig. 23.

The texture of a starch paste is also affected by cross-linking. Starch pastes are classified as being either *long* or *short*. A short paste spoons well (cleanly), whereas a long paste tends to be stringy and does not spoon well. A convenient test for paste texture is shown in Fig. 24; a short paste gives a narrow deflection, and a long paste gives a broad deflection. Cross-linking of a starch makes its paste much shorter than the paste of the unmodified starch. Even low levels of cross-linking decrease the swelling and solubility of the starch. Therefore, the solution contains less starch that can interact with its neighbor (the soluble fraction) to produce a long paste.

Many starch gels, particularly when stored under cool conditions, become opaque with time. The phenomenon is caused by retrogradation, i.e., the starch crystallizing. Crystallization is faster if the chains are smaller and more mobile. Therefore, cross-linking delays retrogradation and thereby delays the time when the starch gel becomes opaque.

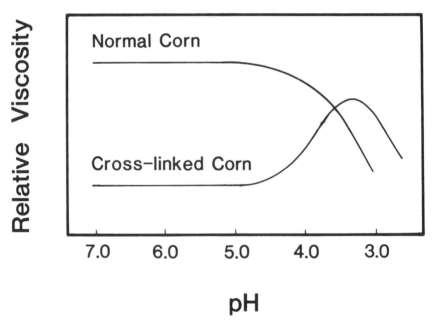

Fig. 23. Effect of pH on the viscosity of normal and cross-linked normal corn starches.

A similar phenomenon is found when gels are frozen and thawed. During freezing, the starch chains are rendered immobile and ice crystals are formed. During storage, even at subfreezing temperatures, water migrates from small crystals to large ones. As a result, crystals are fewer but larger. Because the water is now concentrated into certain areas, the product also has voids. As the ice thaws, the first water produced supplies the starch chains a solvent and thus mobility. In unmodified starch, the chains rapidly interact with each other; the water released during additional melting cannot penetrate the interaction, and the product loses its gel consistency. The gel becomes opaque and watery and has a tough rubbery texture. Cross-linking holds the starch chains fixed in space so they cannot interact strongly. Therefore, the water produced during melting can again hydrate the starch chains, and the starch gel retains its properties. Cross-linked starch can withstand a number of freeze-thaw cycles.

SUBSTITUTED

If we form a monoester of phosphoric acid on starch, the starch is *substituted* rather than cross-linked. We have added a bulky group as well as one that is charged. Both of these tend to make the starch

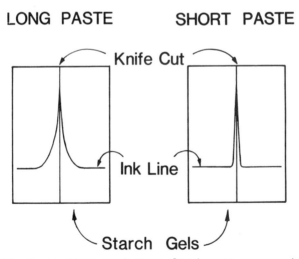

Fig. 24. Test for short and long starch pastes. Starch pastes are poured and allowed to cool. An ink line is drawn across the gel, and then the gel is cut with a knife at right angles to the ink line. With a long paste, the line is distorted futher from the knife cut, showing the interaction of the molecules. With a short paste, the knife does not cause much distortion.

chains repel each other. Thus, the granule tends to swell and solubilize more during gelatinization. This gives a starch paste with higher viscosity but with poorer resistance to shear thinning. Because the chains tend to repel each other, they do not interact or crystallize as easily as those in native starch. Thus, substitution stops retrogradation and opaqueness of starch gels and is helpful in giving improved *freeze-thaw stability*. High levels of substitution (~0.7 DS of phosphorylation) give starches that gelatinize at room temperature. This type of modified starch is useful in making instant pie fillings and puddings. A number of chemical groups can be grafted onto starch to give a substituted starch.

Starch gels can be produced with a wide range of properties by varying the source of the starch as well as the degree and type of modification. One can use a mixture of unmodified, cross-linked, or substituted starches to achieve the desired viscosity, paste clarity, and freeze-thaw stability desired for a specific product.

OTHER MODIFIED STARCHES

Large quantities of *oxidized starches* (generally treated with hypochlorite) are produced. They are used principally in nonfood applications, i.e., as laundry starch and for paper manufacture. They are also used in breading formulations and are reported to give improved adhesion to meat products.

Although not strictly a modified starch (no covalent bond is formed), starch clathrates are also useful in modifying the properties of starch gels. The clathrate forms, presumably, with amylose and retards both swelling and solubility of starch. The reduced solubility results in a much shorter gel. An example is the use of monoglycerides to form a clathrate that gives a much fluffier instant mashed potato product.

Another modified starch that contains no new covalent bonds is *cold-water-swelling starch*. Such starches are produced by heating native starch granules in an alcohol-water mixture. The starch crystallites are melted, but the water is insufficient to swell the starch granules. The solvent is then removed, and the dried starch is stable. Addition of these granules to cold water results in their swelling. Thus, with these products, foods can be thickened without heat except for that contained in the cold water.

STARCH-DERIVED FAT MIMICS

Recently there has been great interest in producing no-fat or reduced-fat foods. Many of the ingredients used to produce such foods are based on starch.

The role of fat in foods is complex and not well understood. Several of its important roles are to act as a plasticizer, to help incorporate air, to soften the mouth-feel of foods (often referred to as adding moistness), and to trap flavors. These roles, except for trapping flavors and incorporating air, can be compensated for by higher levels of water in the product.

Thus, we can visualize the role of the starch-based fat mimics as poducing products with much higher water contents. To do this and still produce a reasonable product requires that the water be stabilized in some way. There are essentially three types of starch-based mimics. Long-chain starch molecules control water by acting like many of the gums that are also used as fat mimics. The major problem with this type is that they interact with themselves to form strong gels or to recrystallize (retrograde). At higher concentrations, short-chain dextrins also control water and thus act as fat mimics. If the viscosity is not high enough, these also may crystallize. The third type is the microcrystalline particles produced from sheared, acid-hydrolyzed starch. These particles control water and give the desired product.

None of the starch-based mimcs, by themselves, incorporate air or trap flavors. However, they do have interesting properties and will find many uses in foods, some of which will have nothing to do with replacing fat.

Conversion of Starch to Sweeteners

Because starch is composed essentially of glucose, hydrolysis produces glucose syrup. Large amounts of starch are commercially converted to syrups. Both the α-1,4 and α-1,6 bonds in starch are susceptible to acid hydrolysis. It would appear straightforward to cook starch with acid to produce syrup. However, numerous side reactions can and do occur. From a practical standpoint, acid hydrolysis is effective only to thin the starch (reduce its viscosity) as it is being gelatinized.

Before continuing with a discussion of hydrolyzing starch to syrup, we should discuss sweeteners and sweetness. The relative sweetness of a number of sugars is given in Table II. Sucrose is taken as the reference and given a value of 100. Great variation exists in the way different people perceive sweetness. What is very sweet to me may be only slightly sweet to you. Also, the relationship between the concentration of sugar and the perceived sweetness is not linear. Sweetness perception is also affected by pH, as well as by other materials in the food. However, the values in the table, although not absolute, are still useful. Glucose is only about 70% as sweet as sucrose, and maltose is only 30% as sweet. Sugars the size of maltotriose and larger are not sweet. Therefore, to produce a sweet syrup from starch, we must convert much of the starch to glucose. It is also of interest to note the high relative sweetness of fructose.

Another concept important in understanding conversion of starch to sugars is that of *dextrose equivalent* (DE). *Dextrose*, the trivial name for glucose, is used extensively in the wet-milling industry. Glucose is a reducing sugar because it has a potentially free aldehyde group on carbon-1. If the oxygen on C-1 is linked to another molecule, as in an α-1,4 or α-1,6 link, then the glucose residue is no longer reducing. Thus, free glucose is a reducing sugar, whereas in maltose one of the glucose molecules is reducing and the second one is not. In starch, there is only one reducing group in each starch molecule; all of the other glucose molecules are linked at the C-1 position. Therefore with each hydrolytic cleavage of an α-1,4 or α-1,6 bond, one reducing group on a glucose molecule is freed. Dextrose equivalent is a measure of the percentage of glucosidic bonds that are hydrolyzed. Complete hydrolysis produces glucose. Thus, if we measure the reducing power of a glucose solution and divide by the total weight of carbohydrate (amount of glucose), we can establish that value as 100 DE, i.e., a value equivalent to 100% dextrose. If we take a glucose chain made up of 10 glucose units linked α-1,4 or α-1,6, measure the reducing power, and divide by the weight of carbohydrate, we will have 10% of the value obtained with pure glucose (a DE of 10). If we hydrolyze one additional bond at any location in the chain, the reducing power doubles but the total carbohydrate stays the same, giving 20% of the value obtained with glucose or 20 DE. Thus, DE tells what percentage of the bonds are broken but does not tell the chemical composition of the resultant syrup.

To produce starch syrup, the starch is gelatinized in the presence of acid. This reduces the viscosity of the starch paste so that large

TABLE II
Relative Sweetness of Several Common Sugars

Sugar	Relative Sweetness
Sucrose	100
Glucose	70
Maltose	30
Fructose	180
Invert syrup	130
High-fructose[a] corn syrup	100

[a] 42% fructose.

amounts of water are not necessary to make it pumpable. Syrups can be made with acid hydrolysis alone; however, at about 40 DE, side reactions start to be important and dark-colored (undesirable) syrups are obtained.

After acid thinning, a number of enzymes can be used for further hydrolysis, depending on the syrup desired. Low-DE syrups, called *dextrins*, that are useful as viscosity builders and as humectants can be made with acid or a combination of acid and α-amylase. Solids having DEs varying from 10 to 25 are sold commercially. They are also useful as flavor diluents. Mixtures of α- and β-amylase are used after acid thinning to produce a high-maltose syrup with a DE of about 42. An alternate path to a high-maltose syrup is to use a debranching enzyme such as pullulanase together with β-amylase. This gives a higher percentage of maltose and a slightly higher DE. Note that a pure maltose solution has a DE of only 50.

Treatment of thinned starch with α- and β-amylase gives syrups with DEs of about 70 at complete reaction. The mixture of α- and β-amylase cannot break α-1,6 bonds; thus, a group of stubs containing α-1,6 bonds is left.

To produce high-DE syrups, glucoamylase must be used. That enzyme produces glucose from the nonreducing end of the starch chain and can hydrolyze both α-1,4 and α-1,6 bonds. Thus, it can theoretically produce a syrup of 100 DE. In commercial practice, values of 92–95 DE are more common. High-DE syrups do contain high levels of glucose and thus are relatively sweet. They are nearly completely fermentable

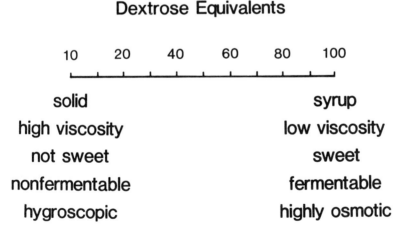

Fig. 25. Properties of various corn syrups.

by yeast and give solutions of high osmotic pressure. The properties of various corn syrups are shown in Fig. 25.

To obtain higher sweetness, part of the glucose must be converted to fructose. This is accomplished with the enzyme *glucose isomerase*. High-DE corn syrup treated with glucose isomerase to equilibrium typically analyzes as 50% glucose, 42% fructose, and 8% higher sugars. This is the high-fructose corn syrup of commerce, which, on a solids basis, is just as sweet as sucrose. This syrup has been very successful in competing with sucrose in many applications. Although it is as sweet as sucrose, it has a higher osmotic pressure and gives a higher water activity than does a sucrose solution. Both glucose and fructose are reducing sugars and therefore more subject to browning than is sucrose, which is not a reducing sugar.

Corn syrups with higher levels of fructose (60 and 90%) have recently been introduced. These are produced by separating the 42% fructose from the mixture by ion-exchange techniques.

REVIEW QUESTIONS

1. What three cereals have starch granules that vary both in size and shape?
2. What is a lenticular starch granule?
3. As what compound is carbohydrate transported from the leaves to the grain in most plants?
4. What is meant by the statement that starch grows by apposition?
5. Starches give three types of X-ray diffraction patterns. What are they? Give an example of a starch that gives each pattern.
6. A fourth type of X-ray diffraction pattern found with starch is the V-pattern. What causes this pattern?
7. What is birefringence?
8. What are the differences between amylose and amylopectin?
9. What are clathrates?
10. What is meant by A-, B-, and C-chains in amylopectin?
11. What are the action patterns of α-amylase and β-amylase?
12. What do the enzymes pullulanase and isoamylase do?

13. How can one separate starch into amylose and amylopectin?
14. Which starch polymer is responsible for the crystallinity in the starch granule?
15. How is amylopectin thought to be arranged in the starch granule?
16. What is a waxy starch?
17. What is gelatinization?
18. What is pasting?
19. What is responsible for setback?
20. What is a starch gel?
21. What is syneresis?
22. What is retrogradation?
23. During starch gelatinization in an excess of water, birefringence is lost over a temperature range of about how many degrees Celsius? How does this change in limited water?
24. How are Naegeli starches produced?
25. What are the properties of acid-modified starch?
26. What reagents are commonly used to cross-link starch?
27. What is degree of substitution (DS), and what DS would be used in food starches?
28. How does cross-linking affect the properties of starch?
29. What are long and short starch pastes?
30. What is a substituted starch?
31. How are cold-water-swelling starches made?
32. What are fat mimics?
33. What are the three types of starch-based fat mimics?
34. Rank the following sugars in order of decreasing sweetness: glucose, fructose, sucrose, maltose.
35. Define dextrose equivalent.
36. What is glucoamylase, and why is it important in producing corn syrups?

SUGGESTED READING

BANKS, W., and GREENWOOD, C. T. 1975. Starch and its components. Edinburgh University Press, Edinburgh.

D'APPOLONIA, B. L., GILLES, K. A., OSMAN, E. M., and POMERANZ, Y. 1978. Carbohydrates. Pages 301-392 in: Wheat Chemistry and Technology, 2nd ed. Y. Pomeranz, ed. Am. Assoc. Cereal Chem., St. Paul, MN.

DENGATE, H. N. 1984. Swelling, pasting, and gelling of wheat starch. Pages 49-82 in: Advances in Cereal Science and Technology, Vol. 6. Y. Pomeranz, ed. Am. Assoc. Cereal Chem., St. Paul, MN.

DONOVAN, J. W. 1979. Phase transitions of the starch-water system. Biopolymers 18:263-275.

EVANS, I. D., and HAISMAN, D. R. 1982. The effect of solutes on the gelatinization temperature range of potato starch. Starch/Staerke 34:224-231.

FRENCH, D. 1984. Organization of starch granules. Pages 184-247 in: Starch Chemistry and Technology, 2nd ed. R. L. Whistler, J. N. BeMiller, and E. F. Paschall, eds. Academic Press, Orlando.

GREENWOOD, C. T. 1976. Starch. Pages 119-157 in: Advances in Cereal Science and Technology, Vol. 1. Y. Pomeranz, ed. Am. Assoc. Cereal Chem., St. Paul, MN.

LINEBACK, D. R., and RASPER, V. F. 1988. Wheat carbohydrates. Pages 277-372 in: Wheat: Chemistry and Technology, 3rd ed. Vol. 1. Y. Pomeranz, ed. Am. Assoc. Cereal Chem., St. Paul, MN.

MANNERS, D. J. 1985. Some aspects of the structure of starch. Cereal Foods World 30:461-467.

WARZBURG, O. B. 1986. Modified Starches: Properties and Uses. CRC Press, Boca Raton, FL.

ZOBEL, H. F. 1984. Gelatinization of starch and mechanical properties of starch pastes. Pages 285-309 in: Starch Chemistry and Technology, 2nd ed. R. L. Whistler, J. N. BeMiller, and E. F. Paschall, eds. Academic Press, Orlando.

Proteins of Cereals

CHAPTER 3

Proteins are naturally occurring polymers found in all living organisms. They are composed of amino acids linked together by peptide bonds.

Structure

The structures of the common amino acids found in proteins can be found in any general biochemistry book. As the name implies, all amino acids have an acid group and an amino group, but they vary in the structure of the R group (the part of the amino acid not involved in the peptide bond). Amino acids are commonly grouped by their type of R group. Four common groups are given in Fig. 1.

Proteins vary in molecular weight from a few thousand to several million. A protein of with a molecular weight of 100,000 would contain about 850 amino acid residues.

The acidic and amino groups of each amino acid are involved in the peptide bonds and form the backbone of the protein. The backbone structure of all proteins is essentially the same. The primary structure, i.e., the sequence of amino acids, is the first level of differentiation between proteins; however, for most of the functional differences in proteins one must look to the secondary and tertiary structures. The peptide bonds that make up the backbone of the protein are flexible to a limited extent and can twist or curl the polypeptide into different forms. The sulfhydryl group on the amino acid cysteine is an active group; it can react with another cysteine residue to form a disulfide bond (-S-S-). That linkage is one factor that gives the protein its secondary structure. The two cysteine residues can be on the same protein chain (intramolecular bonding), forming a loop in the protein, or they can be on different protein chains (intermolecular bonding), linking two peptide chains together (Fig. 2).

A number of different types of noncovalent bonds are responsible for the tertiary structure of proteins. In most cases, the individual bonds are relatively weak, but the large number of them creates overall strength and stabilizes the structure. Two examples of these bonds are ionic bonds (salt formation between an acidic and a basic group) and hydrogen bonds (which are very prevalent) with side chains containing uncharged oxygen, nitrogen, and hydrogen (Fig. 3). Another type of bonding is hydrophobic bonding—two hydrophobic side chains associate closely with each other and, because of Van der Waal and other forces, form a relatively strong bond. When a protein is placed in solution, a number of forces are active. For example, positive charges repel other positive charges and attract negative charges. Hydrophilic groups associate together, as do hydrophobic groups, and each repels the other type. The sum of all this activity determines the tertiary structure of the protein. However, the final determinant of how the protein will fold or curl to obtain its final three-dimensional structure is the sequence of the amino acids.

It is the three-dimensional structure of the protein that determines its properties. Whether or not it is soluble in water depends on a number of factors: for example, its molecular weight (larger molecules are generally less soluble) and whether charges and other hydrophilic groups are on the outside of the molecule where they can interact with water or are buried in the interior of the molecule. Whether or not the protein has enzymatic activity is determined by the structure of the enzyme and how, or if, it binds a substrate molecule.

ACIDIC	BASIC	NEUTRAL (HYDROPHILIC)	NEUTRAL (HYDROPHOBIC)
GLUTAMIC ACID	LYSINE	GLUTAMINE	VALINE
ASPARTIC ACID	HISTIDINE	ASPARAGINE	LEUCINE
	ARGININE	SERINE	ISOLEUCINE
	TRYPTOPHAN	THREONINE	ALANINE
			PHENYLALANINE
			TYROSINE
			CYSTEINE
			CYSTINE
			PROLINE
			METHIONINE
			GLYCINE

Fig. 1. Grouping of amino acids based on their charge and hydrophobicity.

When the three-dimensional structure of the protein is destroyed or altered, the protein is said to be *denatured*. In essence, the weak tertiary bonds are broken, and the protein goes to a random structure. Of course, this changes the physical properties of the protein. For example, enzyme activity is lost. A classic example of denaturation occurs when protein is heated in an aqueous medium. The system's kinetic energy, which increases as the temperature is increased, breaks hydrogen bonds, and the protein goes from its original configuration to a more random one. Alteration of pH and treatment with various reagents also cause protein to be denatured. Proteins usually become less soluble as a result of denaturation, although, like most rules, this generalization is not always true.

Denaturation is usually a nonreversible event. However, it can be reversed in some cases. We can rationalize this behavior by considering that, when the denaturing force or reagent is removed, the protein assumes its most thermodynamically favored configuration. If that is the same as the original configuration, then the protein is renatured.

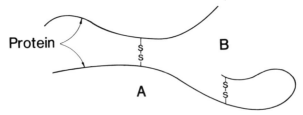

Fig. 2. Illustration of disulfide bonding between polypeptide chains (A) and within a polypeptide chain (B).

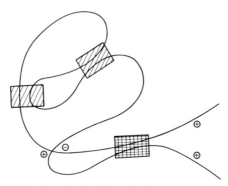

Fig. 3. Illustration of ionic bonds (positive, basic side chains and negative, acidic side chains), hydrogen bonds (diagonal lines show, for example, glutamine), and hydrophobic bonds (hatched lines show, for example, leucine) in proteins.

Proteins in solution are affected to a large extent by both the pH and the ionic strength of the solution. These effects are brought about by the charge on the protein molecule and by how much the charge is shielded. As the pH of the medium is changed, the charge on the protein changes either in sign or in intensity. Because these charges change, the three-dimensional structure of the protein also changes. This is the rationale for why pH changes the activity of enzymes. Salt can shield the charges on a protein molecule by becoming ordered around the charge. Such action negates the charge's effect on the protein structure.

Classification of Proteins

Traditionally, proteins have been classified into four types according to their solubility. This classification is based on the classical work of T. B. Osborne early in the 20th century.

Albumins are proteins soluble in water. Their solubility is not affected by reasonable salt concentrations. In addition, these proteins are coagulated by heat. The classic example of this type of protein is ovalbumin (egg white).

Globulins are proteins insoluble in pure water but soluble in dilute salt solutions and insoluble at high salt concentrations. This class of proteins shows the classic salting in and salting out.

Prolamins are proteins soluble in 70% ethyl alcohol.

Glutelins are proteins soluble in dilute acids or bases.

This type of classification based on solubility is used because it works and has stood the test of time. It gives reproducible results that tell something about the protein. However, the fractions obtained are not clear-cut. For example, prolamins have limited solubility in water, particularly at low ionic strength. In general, each group has subgroups, and certainly none of the groups consists of a single pure protein.

There are also proteins that do not appear to fall into any of the four groups. Wheat, barley, and rye contain glycoproteins soluble in water but not coagulated by heat. Corn, sorghum, and rice have proteins not solubilized by dilute acids or bases. Are these proteins a special subgroup of glutelins or a separate group? Clearly, better classification schemes and methods are required. However, all of those suggested to date become so complicated that they quickly become useless. Therefore, it appears better to use solubility as the first classification scheme and then to rely on electrophoresis, isoelectric focusing, gel filtration, and other analytical separating techniques to further identify proteins.

Properties of the Protein Solubility Groups

Most of the physiologically active proteins (enzymes) are found in the albumin or globulin groups. In cereals, the albumins and globulins are concentrated in the aleurone cells, bran, and germ, with somewhat lower concentrations in the endosperm. Nutritionally, the albumins and globulins have a very good amino acid balance. They are relatively high in lysine, tryptophan, and methionine, three amino acids that are found in relatively low amounts in cereals.

The prolamins and glutelins are the storage proteins in cereals. The plant stores protein in that form for use by the seedling during germination. These proteins are limited to the endosperm in cereals and are not found in the pericarp or germ. The prolamins of all cereals are low in the nutritionally important amino acids lysine, tryptophan, and methionine. The glutelins appear to be more variable in amino acid composition. In wheat, the composition of the glutelins is similar to that of the prolamins. However, this is not true in corn. The glutelins of corn have a much higher lysine content than do the prolamins.

In corn, at least, and probably in the other cereals, the ratio of the various protein groups is under genetic control. Certain varieties of corn have different ratios of protein. For example, "high-lysine" corn mutants have two to three times as much albumin and globulin, one third as much prolamin, and much higher levels of glutelins than those of normal corn. Because the prolamins are low in lysine, the net result of these differences is a corn cultivar with a much higher lysine content.

The storage proteins of wheat are unique because they are also functional proteins. They do not have enzyme activity, but they function to form a dough that will retain gas and produce light baked products. This function is discussed in Chapter 12.

Variation in Protein Content

Cereal grains, like all biological materials, vary widely in their chemical composition. The variation in composition is quite noticeable in the protein content. Wheat varies from less than 6% to more than 27% protein, although most commercial samples are between 8 and 16% protein. The wide variation is the result of both environmental and genetic effects. Protein is synthesized throughout the fruiting period of the plant. Starch synthesis, on the other hand, starts later during fruiting and accelerates as maturity approaches. Thus, if growing conditions late in the fruiting period are good (providing adequate moisture and nutrients), starch yield will be good and grain yields high, but protein contents will be relatively low. Of major importance, of course, is the availability of nitrogen throughout the growing period. An excess of nitrogen early in the growth cycle produces an increase in yield, whereas an excess of nitrogen later in the cycle (after flowering) results in increased protein content.

Other environmental factors that might result in high protein are drought (particularly late in fruiting), frost damage, and certain diseases. Both the frost and the diseases can stop the normal deposition of starch and other constituents, including protein; however, the net result is a higher protein content.

Until relatively recently, protein content in cereal grains was thought to be controlled only by environmental factors. However, certain cultivars of wheat were found to consistently produce grain of higher protein content than that of control cultivars. This led to studies showing that wheat protein content is also under genetic control. Therefore, a plant breeder can breed for higher (or lower) protein content just as for other attributes.

The protein content of cereals is important for two reasons. First, protein is an important nutrient in our diet. Thus, the type and amount of protein is important from a nutritional standpoint. Second, the amount and type of protein is important in the functional uses of the flour. For example, protein content is probably the most important factor in bread flour quality.

When the protein content of a cereal grain changes, the relative proportions of the various proteins also change. This can be seen in Fig. 4. At a low protein content, the amount of albumins and globulins, expressed as a percentage of the total protein, is much higher than at a higher protein content. The total amount of albumins and globulins increases as the amount of protein in the sample increases. But as a percentage of the total protein, it does not increase as fast as the percentage of storage proteins. This appears logical if we remember that the albumins and globulins are physiologically active proteins and the prolamins and glutelins are storage proteins. As the plant produces more protein, less is required for physiological functions and more is available as storage protein.

The nutritional implication of the above is obvious if we remember that albumins and globulins have good amino acid compositions and that the storage proteins are, in general, poorer in amino acid composition.

Wheat Proteins

Among the cereal flours, only wheat flour has the ability to form a strong, cohesive dough that retains gas and produces a light, aerated baked product. Wheat proteins, and more specifically the gluten proteins, are believed to be primarily responsible for that uniqueness of wheat. Because of that important uniqueness, the gluten proteins are discussed separately (in Chapter 10).

Wheat protein synthesis in the grain appears to start soon after anthesis. Gluten proteins have been reported as early as six days after flowering. In the early stages of protein deposition, the storage proteins are deposited into protein bodies. The bodies can be identified after about 12 days. However, as the grain matures, and particularly as it dries, the protein bodies appear to lose their distinct structure and are no longer distinguishable. The protein becomes the well-known, if misnamed, amorphous matrix.

In the process of milling wheat to white flour, the total protein content of the flour decreases an average of about 1%. Considering that the

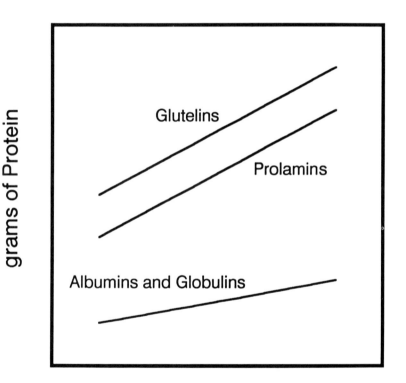

Fig. 4. Change in protein groups as percentages of total protein as protein content changes.

flour makes up slightly more than 70% of the total wheat, this shows that the mill-feed fraction is much higher in protein than the endosperm is. The proteins in the mill feed do not contain any gluten proteins. They are, however, much better from a nutritional point of view. Relatively little is known about the mill-feed proteins other than their amino acid composition.

The amino acid composition of wheat and flour is given in Table I. The flour is higher in glutamic acid and proline, while wheat is higher in lysine, arginine, and aspartic acid.

The proteins in the flour are predominantly the gluten proteins. However, about 15% of the total protein of flour is made up of soluble proteins (albumins and globulins). The amino acid composition of these soluble proteins (Table I) is much different from that of the total flour proteins. In general, the soluble proteins are lower in glutamic acid and proline and much higher in lysine, arginine, and aspartic acid. This pattern is similar to that of the mill-feed proteins. The soluble proteins contain most of the many enzymes found in flour.

Also noticeable in the amino acid composition of flour is the high level of ammonia. This high level comes from the breakdown of

TABLE I
Amino Acid Composition[a] (g/100 g of protein) for Wheat, Flour, and Water- and Salt-Soluble Proteins from Wheat[b]

Amino Acid	Wheat	Flour	Albumins	Globulin
Lysine	2.8	2.0	4.8	5.1
Histidine	2.4	2.1	2.2	3.1
Arginine	4.4	3.2	5.2	10.7
Aspartic acid	4.9	3.8	7.7	8.0
Threonine	2.8	2.6	3.8	3.8
Serine	4.5	4.5	4.0	4.5
Glutamic acid	32.3	35.4	24.6	19.2
Proline	10.6	11.7	9.4	4.8
Glycine	4.0	3.4	4.2	5.4
Alanine	3.5	2.9	5.1	5.2
Cysteine	2.4	2.5	2.8	2.2
Valine	4.2	4.1	6.1	6.5
Methionine	1.2	1.2	1.9	2.0
Isoleucine	3.4	3.6	3.3	3.9
Leucine	6.7	6.7	7.2	7.4
Tyrosine	1.7	1.4	2.6	3.2
Phenylalanine	4.6	4.8	4.9	4.6
Ammonia	3.6	4.0	...	...

[a] Data from F. K. Shoup, Y. Pomeranz, and C. W. Deyoe. 1966. J. Food Sci. 31:94-101 and from S. Dubetz, E. E. Gardiner, D. Flynn, and A. Ian de la Roche. 1979. Can. J. Plant Sci. 59:299-305.
[b] For gluten amino acid composition see Table I, Chapter 10.

glutamine to form glutamic acid during the acid hydrolysis used as part of the analytical procedure. Essentially all of the glutamic acid occurs as glutamine in the native proteins. Because of the high nitrogen content caused by the high level of glutamine in wheat, its protein content is estimated as its nitrogen content times 5.7, rather than the factor of 6.25 used with most other cereals.

Proteins in Other Cereals

The proteins of cereal grains other than wheat do not have dough-forming properties to any extent. Rye and triticale probably come closer than the others, but still their doughs are weak at best. In many parts of the world, the so-called coarse grains (corn, sorghum, and pearl millet) are used to make dough-type products, such as the tortilla of Central and South America or the *roti* or chapati of India. The dough produced is quite different from a wheat flour dough. The major cohesive force appears to be that created by the surface tension of water rather than by the cereal proteins.

The protein content of other cereals is estimated as the nitrogen content times 6.25. This factor is used for all cereals except rice and wheat.

CORN

Proteins occur in the endosperm of corn as discrete protein bodies and a matrix protein (see Chapter 1). The protein bodies are composed mainly of a prolamin called *zein*. The endosperm contains ~5% albumins plus globulins, ~44% zein, and ~28% glutelins. The remaining ~17% is a fraction not found in wheat by the classic Osborne-Mendel fractionation procedure.[1] It is a zein fraction cross-linked by disulfide bonds that is soluble in alcohol containing mercaptoethanol or a similar solvent. The addition of this reducing, alcoholic solvent creates what is basically the Landry-Moureaux procedure[2] of classifying cereal proteins.

The amino acid composition of corn endosperm is given in Table II. Corn proteins have a high level of glutamic acid, but only about half the level found in wheat. The low level of ammonia nitrogen shows that the glutamic acid is present as the acid and not as the amide.

---

[1]This procedure is explained in *The Vegetable Proteins*, 2nd ed. Longmans, Green, and Co., London, 1907.
[2]See Bull. Soc. Chim. Biol. 52:1021-1027 (1970).

Of particular interest in the amino acid composition is the high level of leucine. This has been implicated in the incidence of pellagra (a B vitamin deficiency disease). Both the water- and salt-soluble fractions and the glutelin fraction appear to have reasonably good amino acid balance (Table III). The zein and *cross-linked zein* fractions are low in lysine and very high in leucine. Also of interest is the high level (18%) of proline in the cross-linked zein fraction. Corn, like sorghum and pearl millet, has both vitreous and opaque endosperm in its kernel, and the distribution of protein is different in these two types of endosperm. Because the proteins are different, it follows that the amino acid composition is also different.

SORGHUM

In many ways, the proteins of sorghum are similar to those of corn. The prolamin from sorghum, *kafirin*, resembles zein in amino acid

TABLE II
Amino Acid Composition (g/100 g of protein) of Corn and Sorghum Endosperm Proteins

Amino Acid	Corn Endosperm[a]	Sorghum[b]
Lysine	2.0	2.8
Histidine	2.8	2.3
Ammonia	3.3	...
Arginine	3.8	4.5
Aspartic acid	6.2	7.7
Glutamic acid	21.3	22.8
Threonine	3.5	3.3
Serine	5.2	4.2
Proline	9.7	7.7
Glycine	3.2	3.3
Alanine	8.1	9.2
Valine	4.7	5.1
Cystine	1.8	1.3
Methionine	2.8	1.7
Isoleucine	3.8	4.0
Leucine	14.3	13.2
Tyrosine	5.3	4.3
Phenylalanine	5.3	5.1

[a]Data of E. T. Mertz, O. E. Nelson, L. S. Bates, and O. H. Viron. 1966. Adv. Chem. Ser. 57:228.
[b]Data of Jambunathan and Mertz. 1972. Research progress report on inheritance and improvement of protein quality in *Sorghum bicolor* (L.) Moench. Purdue University, West Lafayette, IN. Adapted from J. S. Wall and J. W. Paulis. 1978. Pages 135-219 in: Advances in Cereal Science and Technology, Vol. 2. Y. Pomeranz, ed. Am. Assoc. Cereal Chem., St. Paul, MN.

composition, and the total amino acid composition of sorghum and corn are very similar (Table II). The major difference between the two grains appears to be in the solubility of the prolamins and in the amount of cross-linked prolamins. Kafirin is not soluble in 70% ethanol at room temperature but is soluble if the temperature is increased to 60°C. Kafirin is soluble in 60% t-butyl alcohol at room temperature.

The amount of cross-linked kafirin in sorghum is about 31%, compared with 17% cross-linked zein in corn. The difference is made up in the prolamin fraction: 17% kafirin in sorghum and 20% zein in corn. As in corn, the prolamins (kafirin and cross-linked kafirin) are very low in lysine.

PEARL MILLET

Pearl millet, like sorghum and corn, is a coarse grain, and the three grains are similar in many respects. The amino acid composition of pearl millet varies quite widely, which may reflect the fact that the grain has not been bred extensively. When determined by the Landry-Moureaux extraction procedure, the distribution of protein in pearl millet endosperm appears to be more similar to that of corn than to that of sorghum. Pearl millet also has a high leucine content, but the

TABLE III
Amino Acid Composition (g/100 g of protein)
of Corn Proteins Extracted with Certain Solvents[a]

Amino Acid	Albumins and Globulins	Zein	Cross-Linked Zein	Glutelins
Lysine	4.18	0.46	0.57	4.38
Histidine	2.38	1.28	6.77	2.52
Ammonia	2.36	2.72	2.23	1.68
Arginine	7.35	2.16	3.46	4.49
Aspartic acid	10.06	5.12	1.73	7.90
Threonine	4.60	2.93	3.86	4.04
Serine	5.23	5.11	4.03	5.15
Glutamic acid	14.70	22.18	23.61	16.70
Proline	5.06	9.84	17.83	6.95
Glycine	6.69	2.02	4.72	4.12
Alanine	7.10	9.01	4.92	7.49
Cysteine/2	3.73	2.27	0.87	0.64
Valine	5.28	3.43	6.07	5.27
Methionine	1.72	0.94	1.63	2.86
Isoleucine	4.25	3.53	2.23	3.97
Leucine	6.50	17.49	10.23	12.09
Tyrosine	3.25	4.54	2.52	4.72
Phenylalanine	3.57	6.11	2.56	5.31

[a]Data from J. L. Robutti, R. C. Hoseney, and C. W. Deyoe. 1974. Cereal Chem. 51:163-172.

ratio of leucine to isoleucine is much lower than in sorghum. The ratio is thought to be important in causing the vitamin deficiency disease pellagra.

OATS

Oats are unique among the cereals in that their amino acid balance is quite good from a nutritional standpoint (Table IV). It compares favorably with the standard protein established by the Food and Agriculture Organization of the United Nations. In addition, the protein content of oat groats is, in general, much higher than that found for other cereals. The relatively good amino acid balance is stable even at a higher protein content, also in contrast to the case in other cereals. So in many respects, oats are clearly superior to the other cereals in nutritional content.

As suggested by the above, the distribution of protein in oats is different than in the other cereals. The alcohol-soluble prolamins of

TABLE IV
Amino Acid Composition (% of protein) of Oats and Certain of Its Fractions[a]

Amino Acid	Total Oats	Groats	Endosperm	FAO[b] Scoring Pattern
Lysine	4.2	4.2	3.7	5.5
Histidine	2.4	2.2	2.2	...
Ammonia	3.3	2.7	2.9	...
Arginine	6.4	6.9	6.6	...
Aspartic acid	9.2	8.9	8.5	...
Threonine	3.3	3.3	3.3	...
Serine	4.0	4.2	4.6	...
Glutamic acid	21.6	23.9	23.6	...
Cysteine	1.7	1.6	2.2	...
Methionine	2.3	2.5	2.4	3.5
Glycine	5.1	4.9	4.7	...
Alanine	5.1	5.0	4.5	...
Valine	5.8	5.3	5.5	5.0
Proline	5.7	4.7	4.6	...
Isoleucine	4.2	3.9	4.2	4.0
Leucine	7.5	7.4	7.8	7.0
Tyrosine	2.6	3.1	3.3	...
Phenylalanine	5.4	5.3	5.6	6.0

[a]Adapted from V. L. Youngs, D. M. Peterson, and C. M. Brown. 1982. Pages 49-105 in: Advances in Cereal Science and Technology, Vol. 5. Y. Pomeranz, ed. Am. Assoc. Cereal Chem., St. Paul, MN. Data of Y. Pomeranz, V. L. Youngs, and G. S. Robbins. 1973. Cereal Chem. 50:702-707.
[b]Food and Agriculture Organization of the United Nations.

oats constitute only 10–15% of the total protein. The predominant class appears to be the globulins (~55%), with glutelins making up about 20–25%. The prolamins from oats are called *avenins*.

RICE

In general, the protein content of rice is lower than that of the other cereals. Protein content in rice is calculated as N × 5.95, lower than the factor employed for other cereals but higher than that for wheat. The amino acid composition is relatively well-balanced, with lysine values of about 3.5% of the total protein. Lysine is still the first limiting amino acid, followed by threonine. The reason why threonine is limiting is not clear, as analysis shows adequate levels. The level of glutamic acid is relatively low (<20%).

The classical Osborne fractionation of the proteins shows that the glutelin (*oryzenin*) is the major fraction, being about 80% of the total protein. The prolamin fraction in rice is quite low (~3–5%). To solubilize the rice protein, a solvent of $0.1N$ sodium hydroxide is generally used. Other solvents, including those containing sulfite or mercaptoethanol, are less effective.

RYE

The amino acid composition of rye proteins is slightly better from a nutritional standpoint than that found for most of the other cereals, oats, of course, being a notable exception. Lysine is higher in rye than in wheat or most other cereals (~3.5% of the protein). Tryptophan is the first limiting amino acid. The glutamic acid content is about 25%, and the leucine content is notably low (~6%).

The reason for this well-balanced amino acid composition is the relatively high level of albumins and globulins in rye. The albumins are about 35% of the total protein and the globulins about 10%. Rye prolamins are about 20% of the total. The glutelins, soluble in dilute acid, are only about 10% of the total. However, about 20% of the protein is not solubilized by the normal Osborne scheme.

TRITICALE

This cross between wheat and rye has a protein distribution similar to that of rye. In general, the water- and salt-soluble proteins are slightly lower than for rye, and the prolamins are slightly higher.

BARLEY

Barley, like most of the other cereal grains, is limiting in the amino acid lysine. Threonine is the second most limiting amino acid.

Most barley is harvested with the hull (lemma and palea) intact. The hull makes up about 10% of the total kernel. In general, the hull is low in protein, but its proteins are relatively high in lysine. The germ proteins are also high in lysine, and the endosperm proteins are lower (~3.2%) but still higher than in many cereals. The endosperm is relatively high in glutamic acid (~35%) and proline (~12%). The glutamic acid is present as the free acid and not as the amide, glutamine.

The prolamins of barley are called *hordein*. Hordein, which makes up about 40% of the barley protein, is very low in lysine. The glutelins and especially the albumins and globulins are relatively high in lysine.

REVIEW QUESTIONS

1. What makes up the backbone structure of protein?
2. What bonds are responsible for the primary, secondary, and tertiary structures of proteins?
3. What determines whether a protein is soluble in water?
4. What is denaturation?
5. Explain how pH and ionic strength affect the properties of proteins.
6. Into what classes have proteins traditionally been classified?
7. What groups of proteins are the storage proteins of cereals?
8. What are several techniques useful in separating or characterizing proteins?
9. What is the factor used to convert nitrogen values to protein content for all cereals except wheat and rice?
10. What are zein and cross-linked zein?
11. What is unique about the oat proteins?
12. What is unique about the distribution of protein in rice?
13. What is hordein?

SUGGESTED READING

BIETZ, J. A. 1986. High-performance liquid chromatography of cereal proteins. Pages 105-170 in: Advances in Cereal Science and Technology, Vol. 8. Y. Pomeranz, ed. Am. Assoc. Cereal Chem., St. Paul, MN.

BUSHUK, W., and LARTER, E. N. 1980. Triticale: Production, chemistry, and technology. Pages 115-157 in: Advances in Cereal Science and Technology, Vol. 3. Y. Pomeranz, ed. Am. Assoc. Cereal Chem., St. Paul, MN.

GARCIA-OLMEDO, F., CARBONERO, P., and JONES, B. L. 1982. Chromosomal locations of genes that control wheat endosperm proteins. Pages 1-47 in: Advances in Cereal Science and Technology, Vol. 5. Y. Pomeranz, ed. Am. Assoc. Cereal Chem., St. Paul, MN.

HOSENEY, R. C., VARRIANO-MARSTON, E., and DENDY, D. A. V. 1981. Sorghum and millets. Pages 71-144 in: Advances in Cereal Science and Technology, Vol. 4. Y. Pomeranz, ed. Am. Assoc. Cereal Chem., St. Paul, MN.

JULIANO, B. O. 1985. Polysaccharides, proteins, and lipids of rice. Page 59-174 in: Rice: Chemistry and Technology, 2nd ed. B. O. Juliano, ed. Am. Assoc. Cereal Chem., St. Paul, MN.

KASARDA, D. D., NIMMO, C. C., and KOHLER, G. O. 1971. Proteins and the amino acid composition of wheat fractions. Pages 227-299 in: Wheat: Chemistry and Technology, 2nd ed. Y. Pomeranz, ed. Am. Assoc. Cereal Chem., St. Paul, MN.

KASARDA, D. D., BERNARDIN, J. E., and NIMMO, C. C. 1976. Wheat proteins. Pages 158-236 in: Advances in Cereal Science and Technology, Vol. 1. Y. Pomeranz, ed. Am. Assoc. Cereal Chem., St. Paul, MN.

LASZTITY, R. 1984. The Chemistry of Cereal Proteins. CRC Press, Boca Raton, FL.

SHEWRY, P. R., and MIFFLIN, B. J. 1985. Seed storage proteins of economically important cereals. Pages 1-83 in: Advances in Cereal Science and Technology, Vol. 7. Y. Pomeranz, ed. Am. Assoc. Cereal Chem., St. Paul, MN.

SIMMONDS, D. H., and CAMPBELL, W. P. 1976. Morphology and chemistry of the rye grain. Pages 63-110 in: Rye: Production, Chemistry, and Technology. W. Bushuk, ed. Am. Assoc. Cereal Chem., St. Paul, MN.

TATHAM, A. S., SHEWRY, P. R., and BELTON, P. S. 1990. Structural studies of cereal prolamins, including wheat gluten. Pages 1-78 in: Advances in Cereal Science and Technology, Vol. 10. Y. Pomeranz, ed. Am. Assoc. Cereal Chem., St. Paul, MN.

WALL, J. S., and PAULIS, J. W. 1978. Corn and sorghum grain proteins. Pages 135-219 in: Advances in Cereal Science and Technology, Vol. 2. Y. Pomeranz, ed. Am. Assoc. Cereal Chem., St. Paul, MN.

WRIGLEY, C. W., and BEITZ, J. A. 1988. Proteins and amino acids. Pages 159-276 in: Wheat: Chemistry and Technology, 3rd ed. Vol. 1. Y. Pomeranz, ed. Am. Assoc. Cereal Chem., St. Paul, MN.

WRIGLEY, C. W., AUTRAN, J. C., and BUSHUK, W. 1982. Identification of cereal varieties by gel electrophoresis of the grain proteins. Pages 211-259 in: Advances in Cereal Science and Technology, Vol. 5. Y. Pomeranz, ed. Am. Assoc. Cereal Chem., St. Paul, MN.

YOUNGS, V. L., PETERSON, D. M., and BROWN, C. M. 1982. Oats. Pages 49-105 in: Advances in Cereal Science and Technology, Vol. 5. Y. Pomeranz, ed. Am. Assoc. Cereal Chem., St. Paul, MN.

Minor Constituents of Cereals

CHAPTER 4

The use of the word "minor" in the title of this chapter does not indicate that the constituents discussed here are of less importance than any others. It indicates a quantitative relationship only.

Nonstarchy Polysaccharides

CELLULOSE

Cellulose is the major structural polysaccharide of plants. It is chemically quite simple, being composed of D-glucose linked β-1,4 (Fig. 1). It is a large polymer, its length apparently being determined by its source and how it is isolated. Because it is unbranched and has essentially a linear configuration, it associates strongly with itself and is quite insoluble. In its native state, cellulose is partially crystalline. The high degree of order and insolubility, together with its β linkage,

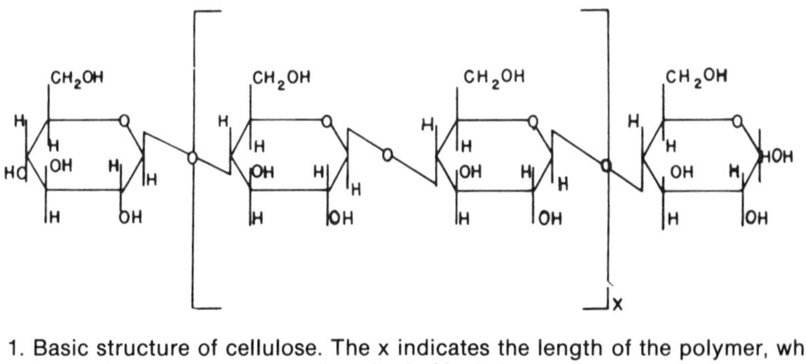

Fig. 1. Basic structure of cellulose. The x indicates the length of the polymer, which varies depending on the source and method of isolation.

makes the polymer resistant to many organisms and enzymes. In plant material, cellulose is usually found associated with lignin and other nonstarchy polysaccharides in cell walls.

Cellulose is a major component of straw, fodder, and hulls; it may make up 40–50% of those plant parts. Thus, those cereals that are harvested with their hulls intact (rice, barley, and oats) contain more cellulose. The pericarp of cereals also is quite rich in cellulose (up to 30%).

Values for the cellulose content of endosperm are usually 0.3% or less. Given the difficulty of determining cellulose quantitatively, this value probably represents no cellulose in the endosperm.

HEMICELLULOSES AND PENTOSANS

The terms *"hemicelluloses"* and *"pentosans"* are often used interchangeably and thus neither appears to have an exact meaning. Taken together, they encompass the nonstarchy and noncellulosic polysaccharides of plants. They are widely distributed in the plant kingdom and, in general, are thought to make up the cell walls and the cementing material that holds cells together (Fig. 2). Chemically, they are quite diverse, varying in composition from a simple sugar such as is found in β-glucans to polymers that may contain pentoses, hexoses, proteins,

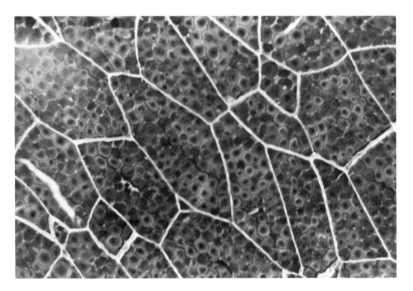

Fig. 2. Photomicrograph showing the cell walls of sorghum.

and phenolics. Sugars that are often reported to be constituents of cereal hemicellulose include D-xylose, L-arabinose, D-galactose, D-glucose, D-glucuronic acid, and 4-O-methyl-D-glucuronic acid.

Much of the confusion concerning the structure of the hemicelluloses probably stems from the difficulty in obtaining a pure chemical entity to study. A compounding problem is the lack of definitive tests to show whether one has a pure entity. Hemicelluloses appear to come in many sizes and with many different chemical compositions.

Another important variation in hemicelluloses is their solubility in water. Wheat flour, for example, contains both water-soluble and water-insoluble hemicelluloses. They are usually referred to in the literature as water-soluble and water-insoluble pentosans, even though they clearly contain many entities in addition to pentose sugars.

The water-insoluble pentosans make up about 2.4% of the wheat endosperm. After removal of gluten by washing, the remaining slurry can be centrifuged to give the water-soluble and water-insoluble fractions. The insolubles can be divided into two layers or fractions based on density. The bottom, more dense layer is the prime starch. On top of this is a gelatinous layer, which has been referred to in the literature by a number of names: amylodextrin, squeegee, sludge, or tailings starch. This fraction is made up of starch, both small-granule and damaged, and water-insoluble pentosans, along with small amounts of proteins and ash. The percentages of each fraction vary widely, depending upon the method of gluten washing and the severity of milling of the flour. The polysaccharides in the water-insoluble pentosans contain L-arabinose, D-xylose, and D-glucose. Reports vary on the presence of galactose and mannose.

Detailed studies of the structure of the water-insoluble pentosans indicate that they are similar to the water-soluble pentosans but have a higher degree of branching. The backbone chain is usually D-xylopyranosyl units linked β-1,4, with L-arabinose side chains. The side chains are characteristically only one sugar residue long. If there is single substitution, the side chain is usually attached to the 3-position of the xylose residue. Substitution at both the 2- and 3-positions can also occur. In those water-insoluble pentosans studied, about 60% of the xylose residues are branched, and about 30% are branched at both the 2- and 3-positions.

When wheat flour is extracted with cold water, 1.0–1.5% soluble pentosans are obtained. Besides the pentoses xylose and arabinose, the water-soluble pentosans usually are reported to contain galactose and protein. Detailed studies of the predominant water-soluble pentosan have shown it to have a backbone of D-xylopyranosyl residue linked β-1,4, with a single anhydro L-arabinofuranosyl residue at the 2- or 3-position (Fig. 3). This side chain appears to be responsible for the solubility of the pentosans, because its removal results in an insoluble xylan polymer. The degree of branching is relatively high, with the arabinose side chains occurring sometimes on isolated xylose residues, sometimes on adjacent residues, less frequently on three residues, but never on four or more consecutive residues.

Early attempts to fractionate the water-soluble pentosans gave many fractions that varied rather widely among studies. This resulted in much confusion. Most of that confusion has now been removed, and the water-soluble pentosans have been shown to consist of an arabinoxylan (contaminated with free protein) insoluble in saturated ammonium sulfate and an arabinogalactan covalently linked to protein. The glycoprotein is soluble in saturated ammonium sulfate.

In addition to the carbohydrates and protein, the water-soluble pentosans contain a small amount of an esterified phenolic acid (ferulic acid, Fig. 4), which is bound only to the soluble arabinoxylans.

OXIDATIVE GELATION

Anyone who has attempted to concentrate flour-water slurries is aware of the fact that water-soluble pentosans form viscous solutions in water. In addition, it has been known for many years that certain oxidizing agents (hydrogen peroxide, for example) cause a large increase in the relative viscosity of a flour-water slurry. In fact, these oxidants

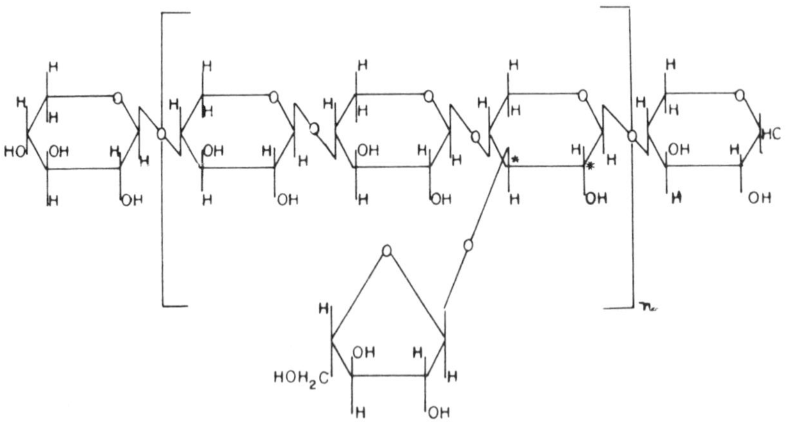

Fig. 3. Structure of the major water-soluble arabinoxylan in wheat flour. n, finite number of polymer units; *, positions at which branching occurs.

can cause the water solubles to gel. The water-soluble pentosans have been shown to be primarily responsible for the gelation. Such oxidative gelation is apparently a unique property of certain cereal flour pentosans, and therefore this property has been studied rather extensively. Because of the loss of ultraviolet absorbance at 320 nm (ferulic acid absorbs at this wavelength) as a result of oxidation, ferulic acid is assumed to be involved in the gelation.

A number of oxidants are effective in causing oxidative gelation. The one that has been studied the most is hydrogen peroxide. Actually, the active agent is a combination of hydrogen peroxide and peroxidase, an enzyme native to flour. Recent work suggests that the oxidative gelation phenomenon is not important in breadmaking. Its importance in the processing of other cereals is not clear.

CEREALS OTHER THAN WHEAT

The hemicelluloses of cereals other than wheat have not been studied extensively. One significant exception to that statement is barley. Because of the function of the cell walls in both malting and brewing, the polysaccharides have been studied in detail. The cell walls of barley contain 70% β-glucan and 20% arabinoxylan, with the remainder being protein and a small amount of mannan. A number of enzymes degrade the cell wall rapidly, particularly after the grain has been germinated. These enzymes have formed the basis for several assays of specific hemicelluloses.

The hemicelluloses (generally referred to as pentosans) of rye have been strongly implicated as a determinant of its baking quality. The amount of pentosans in rye flour is much higher than that found in other cereals, with 8% being a common value. Even with the relatively high concentration and the apparent importance of these pentosans, relatively little has been reported on their chemical composition.

The major hemicellulose in oats is also a β-glucan. As in rye, the level of hemicellulose is higher than that in most cereals, with levels

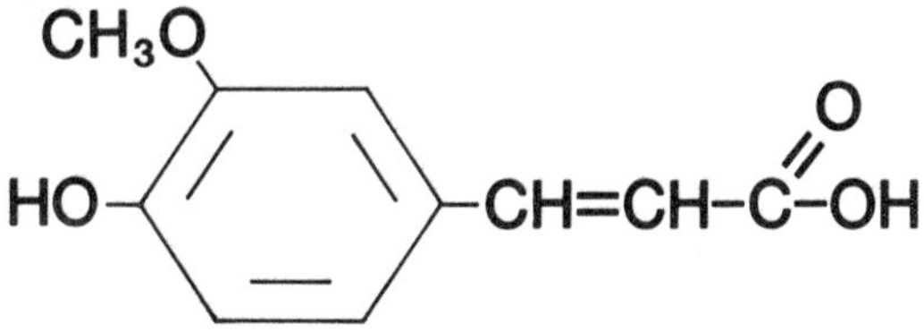

Fig. 4. Structure of ferulic acid.

of β-glucan reported to be 4–6%. β-Glucan has been reported to make up 70–87% of the gums (hemicellulose) of oats. There has been much recent interest in these oat "gums" because of their reported ability to lower cholesterol levels.

When viewed with the scanning electron microscope, the cell walls of the so-called coarse cereals (maize, sorghum, and millets) appear much thinner than the cell walls of wheat or rye (Fig. 13, Chapter 1). The hemicellulose in the walls is much less hydrophilic and does not result in the viscous, slimy mixtures that are common with rye and oats. Chemically, they appear to be rather complicated mixtures containing arabinose, xylose, and glucose as constituent sugars. Rice endosperm also is low in hemicellulose and has thin cell walls. The composition of rice endosperm hemicellulose is a mixture of arabinose-, xylose-, and galactose-containing polymers, as well as protein and a large amount of uronic acids.

Sugars and Oligosaccharides

Sound wheat has been found to contain about 2.8% sugars, including oligosaccharides. The sugars reported include small amounts of glucose (0.09%) and fructose (0.06%), somewhat higher levels of sucrose (0.84%) and raffinose (0.33%), and much higher levels of *glucofructosans* (1.45%). The older literature frequently refers to the presence of maltose (apparently from the breakdown of starch) and to much higher levels of sucrose. The higher sucrose levels were undoubtedly caused by the glucofructosan(s) being misidentified as sucrose.

The glucofructosan(s), also called *levosine*, has sucrose as the smallest member of the series. The next largest is glucodifructose, and the next is an oligosaccharide with the structure shown in Fig. 5. The molecular weight may increase to 2,000 or so. The glucofructosan structure is similar, if not identical, to the structure of inulin, the storage

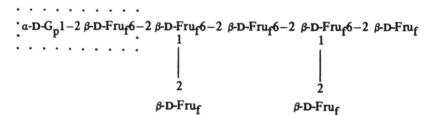

Fig. 5. Structure of a glucofructosan from wheat flour. Dotted lines enclose a sucrose residue.

carbohydrate of some root crops such as the Jerusalem artichoke or chicory.

Sugar content has been used as a monitor to follow the changes that occur during storage of wheat under less than ideal conditions. The complicated changes that occur and the problems associated with detecting the level and type of sugar have made the results difficult to interpret. Recent advances in the analytical methods used for sugar may help these kinds of studies.

Wheat germ contains a rather high level of total sugars (24%), which appear to consist mainly of sucrose and raffinose, with sucrose making up about 60% of the total. No glucofructosans were found in germ that had been hand-dissected from the kernel. Sucrose and raffinose are also the predominant sugars in the bran. The sugars make up 4–6% of the bran. Many other sugars occur in minor amounts in bran. The sugars reported in wheat flour are the same as those reported for the total wheat. The glucofructans appear to be concentrated in the endosperm and absent in germ and bran. One must be very skeptical of any values given in the older literature, as the techniques available before about 1950 did not allow for good separation or identification of sugars.

Brown rice contains about 1.3% sugars. The major sugar is sucrose, with minor amounts of glucose, fructose, and raffinose also reported. Milled rice contains only about 0.5% sugars, and once again sucrose is the major sugar. Oats contain about the same levels of sugars as the other cereals. In the starchy endosperm, sucrose and raffinose are the major sugars. Neither oats nor rice contains any detectable quantity of glucofructosans. In sorghum, the total sugar content may vary from 1 to 6%, the higher sugar levels being in specific cultivars grown for sugar. In those cultivars, sucrose is the major sugar. The trisaccharide raffinose and the tetrasaccharide stachyose are found in small amounts. Values reported for pearl millet are somewhat lower than those for sorghum, ranging from 2.6 to 2.8%. Sucrose accounts for about two thirds of the total. Sorghum and millet apparently do not contain the glucofructosans.

Lipids

The lipids found in cereals are quite complex, mainly because they consist of a large number of chemical classes and a much larger number of individual compounds. The distribution of the classes and compounds is different not only in the various cereals but also in the various anatomical parts of each cereal. Another confusing factor is the fact that lipids can be bound to various other constituents in the cereal, and thus the same chemical entity can exhibit differences in solubility.

Lipids are defined as materials soluble in organic solvents. That is, of course, not a very rigid definition. About the only way to quantitate the amount of lipids is to extract the lipid, make sure it doesn't contain nonlipid species, and determine the amount gravimetrically. The problems with such a scheme are many. Some lipids may be "bound" and require a very polar solvent to extract them from the sample. Polar solvents such as the various alcohols also extract considerable nonlipid material. Thus, the lipids must be redissolved in a nonpolar solvent such as chloroform or petroleum ether before they are weighed. Other lipid entities may be covalently bound to nonlipid materials and must be hydrolyzed before they can be extracted and quantitated. Although acid hydrolysis usually gives the highest values for lipids and is generally considered to give the best results, it also poses problems. For example, lipids that contain a polar entity such as triglycerides are hydrolyzed, but the polar glycerol that results is no longer soluble in lipid solvent and therefore not a lipid by our previous definition. So, although acid hydrolysis may give the largest value, it does not measure all the lipid material that was in the original sample.

With such a complicated picture, how can we understand the lipids? It is really not as complicated as it sounds. One must keep in mind our definition of lipid and the polarity of various solvents. All lipids are soluble in petroleum ether or diethyl ether, both very nonpolar solvents. However, when we extract cereals with such solvents, only part of the lipids is extracted; the rest is bound to proteins, starch, or other entities and therefore is not soluble. We can break many of these associations by using a more polar solvent, methyl alcohol or water-saturated butanol, for example. The ultimate, of course, is acid hydrolysis followed by extraction into a suitable solvent.

To illustrate, if we extract ground whole wheat with ether, a value of about 1.9% extractable lipid is obtained. No purification is necessary, as we have extracted with a relatively nonpolar lipid solvent. Extraction with a highly polar solvent such as water-saturated butanol gives higher values, in the range of 2.2%, but the extract must be purified (dissolved in a lipid solvent) to remove nonlipid material that was soluble in the butanol. Acid hydrolysis gives an even higher value (~2.5%), even though some lipid material is converted into nonlipid material. The only way to obtain a good look at the lipids in cereals is to go through a series of solvents, for example, petroleum ether followed by water-saturated butanol followed by acid hydrolysis. Of course, one can make finer fractionations in lipids by using more solvents that differ in their polarity. In the cereal literature, lipids are often defined as *free* or *bound*; this distinction is based upon solubility. If the lipid is soluble in a nonpolar solvent such as petroleum ether, it is considered free; if it requires a polar solvent for extraction, it is considered bound.

Another important distinction is that of *polar* and *nonpolar* lipids. Actually, only hydrocarbons are truly nonpolar, but this is not the definition used. The common definition of a nonpolar lipid is that material that can be eluted from a silicic acid column with chloroform. By this definition, free fatty acids and triglycerides are nonpolar lipids. The polar lipids are the material eluted from the column with methyl alcohol; this fraction includes the phospholipids and glycolipids.

The distribution of lipids within the wheat kernel varies widely (Table I). Whole wheat lipids contain about 70% nonpolar lipids, 20% glycolipids, and 10% phospholipids. The older literature on lipids must clearly be considered suspect when it is noted that glycolipids (Fig. 6) were not identified in wheat until 1956. The germ has the greatest amount of lipids, and the lipids in the germ have the highest percentage of phospholipids (Fig. 7). The polar lipids of bran contain more phospholipids than glycolipids, whereas the endosperm lipids contain more glycolipids than phospholipids.

Among the many other compounds in the lipids is vitamin E (about 3.9 mg/100 g of whole meal). Wheat lipid contains about 200 mg of total tocopherols per 100 g of oil. The starchy endosperm contains only about 15% of the total tocopherols.

In flour from the starchy endosperm, the lipid content can be divided into the lipids associated with starch granules and the nonstarch lipids. The nonstarch lipids, representing a large number of classes, can be divided into about 60% nonpolar lipids, 25% glycolipids, and 15% phospholipids. The starch-associated lipids also represent a large

TABLE I
Distribution of Crude Fats in Wheat Kernels[a]

Kernel Fraction	Proportion of Kernel (%)	Crude Fat (%)
Whole grain	100	1.8
Bran	15	5.4
Pericarp	7	1.0
Aleurone	6	8.0
Endosperm	82	1.5
Germ	2.5	28.5

[a]Adapted from W. R. Morrison. 1978. Pages 205-348 in: Advances in Cereal Science and Technology, Vol. 2. Y. Pomeranz, ed. Am. Assoc. Cereal Chem., St. Paul, MN.

number of classes, with the following general breakdown: nonpolar lipids (9%), glycolipids (5%), and phospholipids (86%). Clearly, phospholipids make up most of the starch lipids; lysophosphatidylcholine makes up a large percentage (85%) of the phospholipids in starch.

Another variable found in lipids is their fatty acid composition. In general, this is determined more by the species of grain than by the anatomical part of the seed or the lipid type. The starch lipids appear to be an exception to the above rule. The distribution of fatty acids in wheat and its fractions is shown in Table II.

The lipids in barley are reported to constitute about 3.3% of the kernel. About one third of the lipid is in the germ. Since the germ is only about 3% of the total kernel weight, this suggests a lipid content of about 30%. Lipids of the whole barley kernel can be broken down into nonpolar lipids (72%), glycolipids (10%), and phospholipids (21%). Barley also contains tocopherols on the order of 5.0 mg/100 g. The starch-associated lipids in barley are similar to those reported in other cereals. The fatty acids of barley lipids are slightly more saturated than those of wheat.

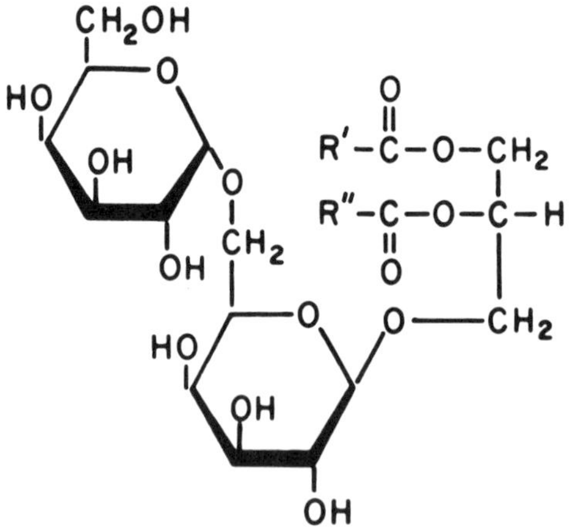

1,2-Diacyl-3-[6-α-D-galactosyl-
β-D-galactosyl]-sn-glycerol
(Digalactosyldiglyceride)

Fig. 6. Structure of digalactosyl diglyceride from wheat flour.

Rye has a lipid content of about 3%, in line with the amounts in barley and wheat. The lipids consist of 71% nonpolar lipids, 20% glycolipids, and 9% phospholipids. Rye germ contains about 12% lipid, somewhat lower than that found in barley or wheat. Rye starch contains about half the level of lipids found in wheat, barley, and oat starches. Lysophosphatidylcholine is the major starch lipid in rye, as in other cereals. Rye contains significantly less $C_{18:2}$ fatty acid than does wheat but significantly more $C_{16:0}$, $C_{18:1}$, and $C_{18:3}$. The high level of $C_{18:3}$ may be responsible for the susceptibility of rye to oxidative deterioration. Triticale, the result of a cross between rye and wheat, has a lipid content and composition similar to those of its parent species.

The lipid content of oats is generally higher than those of the other cereals but also varies quite widely. Values as low as 3% and as high as 12% have been reported. Most lines contain 5–9% lipids. Oats are also unique in that most of the groat lipid (80%) is in the endosperm instead of the germ and bran. Oat lipids have more oleic acid (18:1) than most other cereals. The fatty acids found in oat lipids extracted from the various groat fractions do not differ much. Oats contain only about 2.3 mg of tocopherols per 100 g of grain, somewhat lower than the levels in wheat and barley. However, oats are known for their antioxidant activity. This appears to be the result of a series of phenolic compounds, mainly caffeic and ferulic acid esters of C_{26} and C_{28} 1-alkanols.

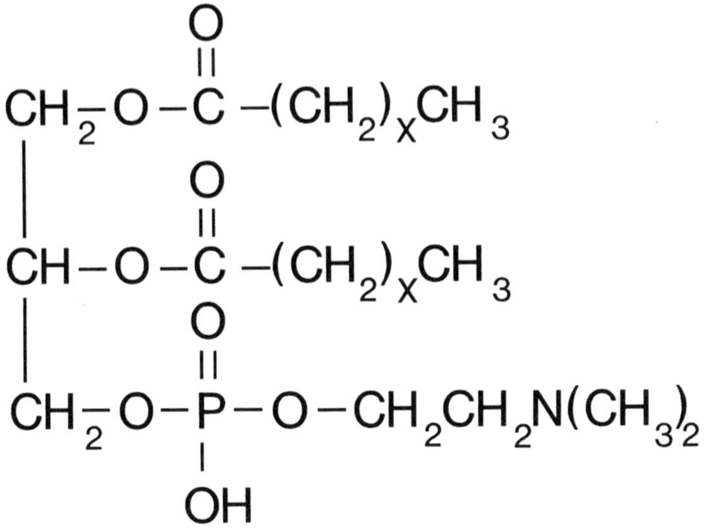

$$\begin{array}{l} CH_2-O-\overset{\overset{O}{\|}}{C}-(CH_2)_xCH_3 \\ | \\ CH-O-\overset{\overset{O}{\|}}{C}-(CH_2)_xCH_3 \\ | \\ CH_2-O-\overset{\overset{O}{\|}}{\underset{\underset{OH}{|}}{P}}-O-CH_2CH_2N(CH_3)_2 \end{array}$$

Fig. 7. Structure of a phospholipid.

Rice contains about 3% lipids, in line with most other cereals. Because the lipid is concentrated in the peripheral parts of the grain, the lipid content decreases as the grain is milled. In fact, fat content has been used as a measure of the degree of rice milling. Milled rice may contain only 0.3–0.5% lipids. Brown rice contains more nonpolar lipids and less glycolipids and phospholipids than do such cereals as barley, wheat, and rye. The fatty acid composition of brown rice is similar to that found in other cereals (Table III). Rice oil also contains a phenolic antioxidant (*oryzanol*), an ester of ferulic acid, and triterpene alcohols.

The oil content of pearl millet ranges generally from 3 to 8%. Reports of the lipid content, like much of the other literature dealing with pearl millet, are confused by the many species of millets and the number

TABLE II
Fatty Acid Composition (%) of Wheat Lipids[a]

Wheat Fraction	Fatty Acid[b]					
	16:0	16:1	18:0	18:1	18:2	18:3
Whole Wheat						
Total	20	1	1	15	57	4
Nonpolar	20	...	...	22	53	3
Polar	18	...	...	15	62	4
Bran	19	1	2	20	50	4
Germ	21	<1	2	13	55	6
Flour						
Nonstarch	19	...	<2	12	63	4
Starch	40	...	<2	11	48	2

[a] Adapted from W. R. Morrison. 1978. Pages 205-348 in: Advances in Cereal Science and Technology, Vol. 2. Y. Pomeranz, ed. Am. Assoc. Cereal Chem., St. Paul, MN.
[b] 16:0 indicates 16 carbons and zero double bonds, etc.

TABLE III
Fatty Acid Composition (%) of Lipids of Various Cereals[a]

Cereal	Fatty Acid				
	16:0	18:0	18:1	18:2	18:3
Corn	13	<4	35	50	<3
Sorghum	12	1	35	49	3
Pearl millet	20	5	25	48	3
Rice	22	<3	39	36	4
Oats	20	2	37	37	4
Rye	18	1	25	46	4
Barley	22	<2	12	57	5
Wheat	21	2	15	58	4

[a] Adapted from W. R. Morrison. 1978. Pages 205-348 in: Advances in Cereal Science and Technology, Vol. 2. Y. Pomeranz, ed. Am. Assoc. Cereal Chem., St. Paul, MN.

of scientific names that pearl millet has had in recent times. The fatty acids of pearl millet free lipids are higher in oleic acid ($C_{18:1}$) than are those of wheat or the other cereals. The quality of pearl millet quickly deteriorates after it has been ground into meal, and the lipid components are generally considered to be responsible for this deterioration. While the mechanism of this deterioration is not clear, it is not the classic oxidative rancidity.

Sorghum contains from 2.1 to about 5.0% lipids. The total lipids are 90% nonpolar lipids, 6% glycolipids, and 4% phospholipids. About 75% of the lipids are in the germ, with the remainder split about evenly between the bran and the endosperm. Sorghum lipids contain about 0.5% wax, which resembles carnauba wax in composition. The fatty acid composition of sorghum lipids (Table III) is similar to that of pearl millet.

Corn is the major cereal used commercially for the production of oil (see Chapter 7). The oil content of corn varies widely. Both the proportion of germ in the kernel and the percentage of oil in the germ vary, and both appear to be under genetic control. In most cultivars, free lipids are about 4.5% and bound lipids somewhat below 1%. The germ makes up a relatively large percentage of the total kernel, about 12% compared to 3% for wheat or barley, and generally contains about 30% lipids. The fatty acid composition of corn lipids is similar to that found in sorghum and pearl millet (Table III).

Enzymes

The cereal grains are complicated biological systems. Even if we restrict our discussion to the fruit or seed, the number of enzymes that are undoubtedly present, at least part of the time, from the initiation of the seed to grain ripeness is essentially unlimited. When viewed from that perspective, essentially all plant metabolic enzymes are present. Reported variations, then, are based on our ability to detect their activity. Therefore, this discussion is limited to the enzymes found at relatively high concentrations and/or those that have been studied extensively.

Because cereals store their excess energy as starch and thus contain relatively high levels of starch, it is not surprising that the starch-degrading enzymes have been studied extensively in cereals. The starch-degrading enzymes are also of importance when cereals are used in malting and brewing (Chapter 9), as well as in breadmaking (Chapter 12).

AMYLASES

Cereals contain two types of amylases. α-Amylase is an endoenzyme that breaks α-1,4 glucosidic bonds on a nearly random basis (Fig. 8). The result of the enzyme's action is, therefore, to rapidly decrease the size of large starch molecules and thereby reduce the viscosity of a starch solution or slurry. The enzyme works much faster on gelatinized starch than on granular starch. However, given sufficient time, it will also degrade granular starch.

Because of its rapid effect on starch paste viscosity, such tests as the amylograph and falling number (both measures of relative viscosity) have been widely used to measure α-amylase activity. A number of direct chemical and enzymatic tests have recently been developed to measure this activity. Sound, intact cereals have low levels of α-amylase. However, upon germination, the level of α-amylase increases many times. This makes α-amylase activity a sensitive measure to detect sprouting of cereal grains. The level of sprouting is generally an important factor in the grains' suitability for food uses.

β-Amylase is an exoenzyme that attacks starch from the nonreducing ends of the polymers (Fig. 9). It also attacks α-1,4 glucosidic bonds and breaks every second bond to release maltose. However, β-amylase will neither break nor bypass α-1,6 bonds. Therefore, the products of β-amylase activity on starch are maltose and *β-limit dextrins*.

One would expect to produce only maltose from amylose, if amylose is unbranched. Actually, only about 70% of the amylose is converted

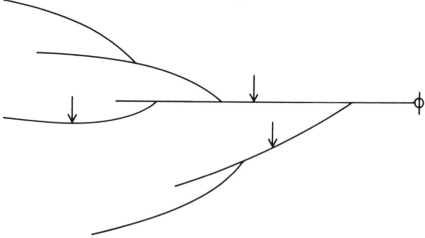

Fig. 8. Action of α-amylase on starch. Lines are glucose units linked α-1,4 and branched α-1,6. ϕ, reducing glucose residue.

to maltose, showing that branching does occur in amylose. With amylopectin, the conversion to maltose is only about 50%, the remainder being a high-molecular-weight β-limit dextrin.

Because β-amylase produces maltose, it is called the *saccharifying* (sugar-producing) enzyme. The combination of α- and β-amylase activity degrades starch quite rapidly and much faster and more completely than either alone. Each break that α-amylase makes in the starch polymer produces a new nonreducing end for β-amylase to attack. β-Amylase alone has practically no action on intact starch granules. Measurement of β-amylase activity is generally difficult, as its "activity," as measured by maltose production, is strongly influenced by the level of α-amylase present. The old "maltose value," sometimes still used in the milling industry, is actually a measure of the action of the combination of those enzymes. A mixture of the two enzymes does not completely degrade starch, as neither enzyme can break the α-1,6 glucosidic bonds present in amylopectin. In fact, the rate of breaking of α-1,4 bonds close to an α-1,6 linkage is extremely low. In general, a combination of the two enzymes results in about 85% conversion of starch to sugar.

Unlike α-amylase, β-amylase is found in sound, intact cereal grains. In general, the level does not increase much as a result of germination. The pH optimum for α-amylase is about 5.3, and that for β-amylase is slightly higher. β-Amylase is slightly more susceptible to heat inactivation than is α-amylase (Fig. 10).

The amylase activity (both α and β) of wheat, barley, and rye appears to be much higher than that found for the other cereal grains. In South Africa, large quantities of sorghum are used in beer production. The measurement of amylase in sorghum malts has been difficult because many of the enzymes are insoluble. Sorghum produces relatively low levels of amylase, but the activity of the purified amylase is similar to that of other cereals. Both α- and β-amylase are produced in sorghum as a result of germination or malting.

Recent work has shown that a relatively large number of amylase isoenzymes are found in cereals. Some are found in specific parts of the grain, others only in immature grains. Also those enzymes produced by germination are different from those natively present in the grain.

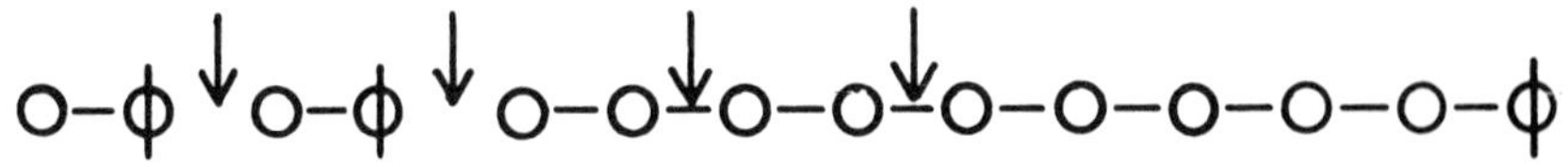

Fig. 9. Action of β-amylase on starch. O, glucose residue; φ, reducing glucose residue.

PROTEASES

Both proteinases and peptidases are found in mature, sound cereals; however, their levels of activity are relatively low. Most methods of determining proteolytic activity are based upon production of soluble nitrogen, which is then measured. With the large and insoluble cereal proteins, significant amounts of enzyme activity may be present, but still no soluble nitrogen may be produced. Recently, methods measuring the endoenzyme activity have been published.

Wheat flour appears to contain a proteolytic enzyme, with a pH optimum near 4.1, that may be of importance in acidic fermentation products such as soda crackers and sourdough bread. The peptidases may be important in producing soluble organic nitrogen that is utilized by yeast during fermentation.

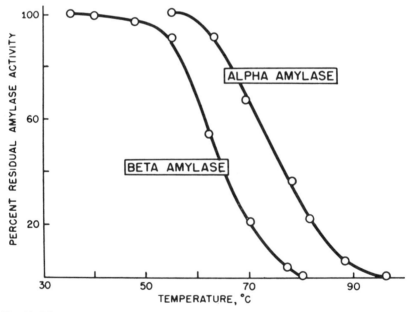

Fig. 10. Effect of temperature on the activity of α- and β-amylases.

LIPASE

Although we generally think of lipase as an enzyme to hydrolyze the fatty acids from triglycerides, it is difficult to separate that activity from the activity of other esterases. All cereals have lipase activity, but the activity varies widely between cereals, with oats and pearl millet having relatively high activity compared to that of wheat or barley. Because of wide variations in results when different techniques or different substrates are used, activities are difficult to compare.

Lipase activity is important because a free fatty acid is more susceptible to oxidative rancidity than is the same fatty acid in a triglyceride. Free fatty acids in a product often give the product a soapy taste.

PHYTASE

Phytase is an esterase that hydrolyzes *phytic acid*. Some question exists as to whether there is a specific phytase or only phosphoesterases. Phytic acid is inositol hexaphosphoric acid (Fig. 11), and the enzyme converts it to inositol and free phosphoric acid. About 70–75% of the phosphorus in cereals occurs as phytic acid, which is thought to chelate divalent cations and keep them from being absorbed in the intestinal tract. Thus, the enzyme's activity appears to be important, as it converts a detrimental entity into inositol (a vitamin) and nutrients. Studies have shown that at least part of the phytic acid in wheat flour is

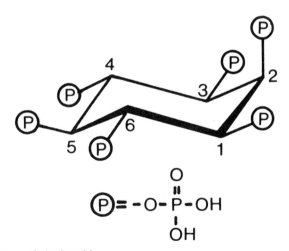

Fig. 11. Structure of phytic acid.

hydrolyzed during fermentation. The solubility of the substrate appears to be the factor limiting hydrolysis.

LIPOXYGENASE

Lipoxygenase catalyzes the peroxidation of polyunsaturated fats by oxygen. The preferred substrate has methylene-interrupted double bonds, with both double bonds in the *cis* configuration. The enzyme is rather widespread in nature. It is in high concentration in soybean but is also found in many other cereals, although it has been reported to be absent in pearl millet. A large number of isoenzymes of lipoxygenase exist and have different activities. One of the major distinctions between them is whether a given isoenzyme will attack the fatty acids on the triglycerides or only the free fatty acids. For example, soybean lipoxygenase attacks triglycerides, whereas wheat lipoxygenase must have free fatty acids to be active. This is why adding enzyme-active soy flour to wheat flour doughs produces enzyme activity.

The enzyme has a number of effects on wheat flour doughs. For one, it is an effective bleaching agent; a coupled oxidation destroys the yellow pigments in wheat flour. This is beneficial in bread dough but a negative factor in pasta products, for which the yellow color is desirable. Most durum wheats have been selected to have low lipoxygenase activity. The enzyme also increases the mixing stability of wheat flour doughs and has been reported to alter dough rheology to produce a strong dough.

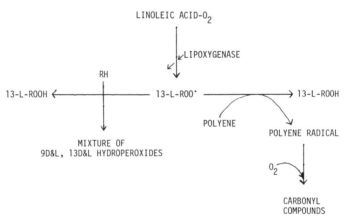

Fig. 12. Competition between linoleic acid (RH) and the polyene crocin for the 13-L-hydroperoxide formed by lipoxygenase. Interaction with crocin results in its bleaching.

A major effect of lipoxygenase is its promotion of the oxidative deterioration of many products. This is brought about by the scheme given in Fig. 12.

OTHER ENZYMES

Most cereals also contain, or produce during germination, enzymes capable of oxidizing o-phenylenediamine. Most of the activity appears to be in the bran. The significance of the enzyme in cereals is not clear. Cereals also contain peroxidase and catalase, both enzymes that use H_2O_2 as a substrate. In comparative studies, peroxidase activity was in varying amounts (wheat > barley ≫ corn > rice). The significance of these enzymes is also not clear.

Vitamins and Minerals

Most cereals are important sources of such vitamins as thiamin, niacin, riboflavin, pyridoxine, pantothenic acid, and tocopherol. In addition, they are good sources of a number of minerals. The vitamin and mineral composition of a number of cereals is given in Table IV. In general, most of the minerals (61% of the total) are concentrated

TABLE IV
Vitamin and Mineral Composition (mg/100 g) for a Number of Cereal Grains[a]

Vitamin or Mineral	Wheat	Rye	Barley	Oats	Rice	Corn	Sorghum
Vitamins							
Thiamin	0.55	0.44	0.57	0.70	0.33	0.44	0.58
Riboflavin	0.13	0.18	0.22	0.18	0.09	0.13	0.17
Niacin	6.4	1.5	6.4	1.8	4.9	2.6	4.8
Pantothenic acid	1.36	0.77	0.73	1.4	1.2	0.70	1.0
Pyridoxine	0.53	0.33	0.33	0.13	0.79	0.57	0.60
Minerals							
Phosphorus	410	380	470	340	285	310	405
Potassium	580	520	630	460	340	330	400
Calcium	60	70	90	95	68	30	20
Magnesium	180	130	140	140	90	140	150
Iron	6	9	6	7	...	2	6
Copper	0.8	0.9	0.9	4	0.3	0.2	0.5
Manganese	5.5	7.5	1.8	5	6	0.6	1.5

[a] Adapted from Simmonds and Campbell (1976) and Hoseney et al (1981). (D. H. Simmonds and W. P. Campbell. Pages 63-110 in: Rye: Production, Chemistry, and Technology. W. Bushuk, ed. R. C. Hoseney, E. Varriano-Marston, and D. A. V. Dendy. Pages 71-144 in: Advances in Cereal Chemistry and Technology, Vol. 4. Y. Pomeranz, ed. Both published by the American Association of Cereal Chemists, St. Paul, MN.)

in the aleurone layer. Vitamins are concentrated in the aleurone or the scutellum or both.

REVIEW QUESTIONS

1. What parts of cereals are high in cellulose?
2. What are pentosans and hemicelluloses?
3. What is the structural difference between water-soluble and water-insoluble pentosans?
4. What is the structure of ferulic acid?
5. What is oxidative gelation?
6. What is the major hemicellulose in oats?
7. What is the typical composition of sugar in wheat flour?
8. What are the glucofructosans?
9. What is the accepted definition of lipids?
10. How are polar and nonpolar lipids defined?
11. In general, do oats have higher or lower lipid contents than other cereals?
12. Commercial oil is obtained from what cereals?
13. What is the action of α-amylase?
14. What are some of the more important enzymes found in cereals?
15. What is the action of lipoxygenase?

SUGGESTED READING

ASPINALL, G. O. 1980. Chemistry of cell wall polysaccharides. Pages 473-500 in: The Biochemistry of Plants, Vol. 3. J. Preiss, ed. Academic Press, Orlando, FL.

BARNES, P. J., ed. 1984. Lipids in Cereal Technology. Academic Press, Orlando, FL.

FINCHER, G. B., and STONE, B. A. 1986. Cell walls and their components in cereal grain technology. Pages 207-295 in: Advances in Cereal Science and Technology, Vol. 8. Y. Pomeranz, ed. Am. Assoc. Cereal Chem., St. Paul, MN.

FOX, P. F., and MULVIHILL, D. M. 1982. Enzymes in wheat, flour, and bread. Pages 107-156 in: Advances in Cereal Science and Technology, Vol. 5. Y. Pomeranz, ed. Am. Assoc. Cereal Chem., St. Paul, MN.

HILL, R. D., and MacGREGOR, A. W. 1988. Cereal α-amylases in grain research and technology. Pages 217-261 in: Advances in Cereal Science and Technology, Vol. 9. Y. Pomeranz, ed. Am. Assoc. Cereal Chem., St. Paul, MN.

HOSENEY, R. C., VARRIANO-MARSTON, E., and DENDY, D. A. V. 1981. Sorghum and millets. Pages 71-144 in: Advances in Cereal Science and Technology, Vol. 4. Y. Pomeranz, ed. Am. Assoc. Cereal Chem., St. Paul, MN.

JULIANO, B. O., and BECHTEL, D. B. 1985. The rice grain and its gross composition. Pages 17-57 in: Rice: Chemistry and Technology, 2nd ed. B. O. Juliano, ed. Am. Assoc. Cereal Chem., St. Paul, MN.

KRUGER, J. E., and REED, G. 1988. Enzymes and color. Pages 441-500 in: Wheat: Chemistry and Technology, 3rd ed. Vol. 1. Y. Pomeranz, ed. Am. Assoc. Cereal Chem., St. Paul, MN.

KRUGER, J. E., LINEBACK, D. R., and STAUFFER, C. E., eds. 1987. Enzymes and Their Role in Cereal Technology. Am. Assoc. Cereal Chem., St. Paul, MN.

LASZTITY, R., and LASZTITY, L. 1990. Phytic acid in cereal technology. Pages 309-371 in: Advances in Cereal Science and Technology, Vol. 10. Y. Pomeranz, ed. Am. Assoc. Cereal Chem., St. Paul, MN.

LINEBACK, D. R., and RASPER, V. F. 1988. Wheat carbohydrates. Pages 277-372 in: Wheat: Chemistry and Technology, 3rd ed. Vol. 1. Y. Pomeranz, ed. Am. Assoc. Cereal Chem., St. Paul, MN.

MORRISON, W. R. 1978. Cereal lipids. Pages 221-348 in: Advances in Cereal Science and Technology, Vol. 2. Y. Pomeranz, ed. Am. Assoc. Cereal Chem., St. Paul, MN.

MORRISON, W. R. 1988. Lipids. Pages 373-439 in: Wheat: Chemistry and Technology, 3rd ed. Vol. 1. Y. Pomeranz, ed. Am. Assoc. Cereal Chem., St. Paul, MN.

WHISKER, E., FELDHEIM, W., POMERANZ, Y., and MEUSER, F. 1985. Dietary fiber in cereals. Pages 169-238 in: Advances in Cereal Science and Technology, Vol. 7. Y. Pomeranz, ed. Am. Assoc. Cereal Chem., St. Paul, MN.

Storage of Cereals

CHAPTER 5

The cereal grains are, in general, amenable to storage for relatively long periods of time. They are usually harvested at a relatively low moisture content and, when stored out of the weather and protected from insects and rodents, easily store for several years. Under ideal storage conditions (low temperature, inert atmosphere, etc.), safe storage may be measured in decades.

Throughout history, the cereal grains have given humans a buffer against crop failure and starvation. In comparison with such foods as dairy products, meats, and fresh vegetables, cereals are relatively easy to store. However, they can and do go "out of condition" if storage conditions are not proper. In the past, and indeed even today in some parts of the world, such loss of cereal stores has led to starvation.

Basic Types of Storage

Grain is generally harvested once or, in some areas of the tropics, twice during the year. Yet it is consumed throughout the year. Therefore, practically all grain *must* be stored. Storage can vary from the simple expedient of pouring the grain on the ground or on streets up to storage in large concrete structures equipped so that a rail car can be picked up and shaken empty in a few minutes.

Generally grain is piled on the ground only during the harvest season when transportation equipment is in short supply. In fact, such storage is not as bad as it sounds. A pile of grain sheds water quite well, and only the top inch or two is damaged with short-term storage. Of course, as storage time increases, the loss increases, because the grain accepts more water from rain and is also exposed to birds, insects, and rodents.

Primitive societies often stored their excess grain underground. The practice continues today in some regions of the world. Underground storage offers a number of advantages. For example, it protects the grain from daily and seasonal variations in temperature; the construction is relatively simple; and it protects grain from insects and molds because of the low oxygen and high CO_2 content of the interseed air. Of course, the site for underground storage must be picked to give a dry environment.

The next level of improvement is storage in bags. Bagged grain can be stored in almost any shelter that protects the bags from the weather and from predators. Bags can be handled without any equipment. However, they are relatively expensive, and handling them is expensive unless labor is very cheap.

Bulk storage in bins is the most widely used type of storage today. The size of the bin may vary from a few hundred bushels for an on-farm storage bin to tens of thousands of bushels for a bin in a terminal elevator. On-farm bins are often constructed of wood or steel, and the larger elevators are today practically all constructed of concrete.

When grain is poured into a bin, it forms an angle from the horizontal that is called the *angle of repose* (Fig. 1). With most grains, this angle is about 27°. Damp grain or very small grain gives a slightly flatter slope. The outflow hopper at the bottom of a bin must be cone-shaped and have a slope greater than the angle of repose, or the grain will not flow out. Smaller bins require a steeper slope because of the greater friction on the sides of the hopper.

The pressure that grain exerts on the bin floor is not proportional to the height of the grain. Because much of the grain weight is supported by the bin's walls, grain follows the laws of semifluids rather than

Fig. 1. Pile of grain, showing angle of repose.

of fluids. Each kernel rests on several kernels below it, so that part of the weight is distributed laterally until it reaches a wall. The lateral pressure of the grain on bin walls is about 0.3–0.6 of the vertical pressure, and the vertical pressure increases very little after a depth of about three times the bin diameter.

Grain settles or packs during storage. Lightweight grain such as oats may pack to lose as much as 28% of its volume. Other, heavier grains may pack only slightly. When grain is poured into a bin, the heavier grain falls faster and straighter. The lighter particles such as chaff accumulate toward the bin walls. However, when the stream of grain hits other grains, small particles are trapped between the larger kernels. Those particles (weed seeds, broken kernels, or heavy dust particles) remain at the pile's center, where the incoming stream hits the grain pile. The whole-grain kernels flow away down the slope (angle of repose). Because the space between kernels may account for 30% of the space in the pile, there may be 30% fines in the area where the grain was introduced. This is called the *spoutline*. When a bin is opened, a column of grain directly above the opening flows out first. The column widens as it reaches further from the opening, giving an inverted cone. The grain in the center flows faster, producing a cone-shaped depression on the surface. The grain at the surface flows into the depression and down to the draw-off.

Moisture, No. 1 for Safe Storage

Unless unusual precautions are taken or unusual conditions occur, all cereal grains contain some moisture. The amount of that moisture depends upon a number of factors and is of paramount importance to anyone having any interest in the grain. As an example, if I have a bushel of wheat that contains no moisture, I have 60 lb of wheat. If I bring that wheat to 15% moisture, I have not 60 but 70.6 lb or 1.18 bu. If wheat is selling for $3.75/bu, my water is now worth $0.66. That is a neat 18% profit. Of course, both buyer and seller are aware of these facts and, therefore, agree on a moisture content for the grain. If they are not aware, they probably will not remain in business very long.

Moisture is also of the greatest importance in the safe storage of grain. Microorganisms, particularly certain species of fungi, are the major cause of grain deterioration. Three major factors control the rate of fungal growth on cereals. These are moisture, time, and temperature. Of the three, moisture is the most important. At low moisture contents, fungi will not grow, but at about 14% or slightly above, fungal growth begins. Between about 14 and 20%, a small addition to the moisture level greatly increases the rate of fungal growth and also changes the number and type of species that develop. Thus, if one is going to store cereals for any period of time, it is important to know the moisture content in any given portion of the stored grain.

A mass of grain in a bin looks deceptively uniform. One can easily assume that the moisture content is also uniform throughout the bin. In fact, this is seldom, if ever, the case. Grain coming from a single field may vary quite widely in moisture content because of differences in the soil or stages of ripeness of the grain. Within a single head or ear of grain, the moisture can, and indeed does, vary from kernel to kernel. If grain is obtained from a number of sources, the moisture will surely vary. We would expect that, given sufficient time, the grain bulk would come to an equilibrium. In fact, this will occur only if the grain is stored under stable conditions, and, in practice, it does not occur, as illustrated in Fig. 2. In addition, other forces, which are discussed later, often upset the equilibrium.

The measurement of moisture content is, at best, difficult. To be completely accurate, we must measure water, rather than other volatile substances, and therefore we cannot simply measure loss of weight. This implies use of the Karl Fischer reagent or something similar. Another factor that is always of great importance is how samples are taken for the analysis. At first glance, obtaining a uniform sample appears relatively simple; in fact, it is extremely difficult. A uniform or average sample may be important if we are buying or selling grain. However, it is of little or no value if we are interested in how the grain will store. The important moisture level is not the average but the *highest* moisture found in our lot.

If one area in the grain has a high moisture content, microorganisms will grow at that point. As the organisms grow, they produce both moisture and heat, as a result of their metabolism, which will then lead to greater damage.

The moisture in grain is in equilibrium with the air surrounding the grain. This equilibrium moisture content (EMC) is defined as the moisture content at equilibrium with an atmosphere at a certain relative humidity. Different lots of grain, even of the same type of grain, may have different EMCs, and various grains in a mixture may have different moisture contents even though they are all at equilibrium with the relative humidity of the interseed air (Fig. 3).

In addition, grains of the same lot in air of the same relative humidity may have different moisture contents, depending upon whether each grain is gaining or losing moisture. This phenomenon, called *hysteresis*, is shown in Fig. 4. The exact reason for the hysteresis effect is not known, but it is accepted as the norm for cereal grains.

The safe storage moisture content of cereal grains is almost completely dependent upon the grain's hygroscopic properties. When in storage, the grain and the moisture content of the associated air come into equilibrium. One of the most damaging factors in grain storage is the growth of molds. Generally molds will not grow on grain in equilibrium with air of less than about 70% rh. The maximum moisture levels for safe storage of the major grains are generally accepted to be: corn, 13%; wheat, 14%; barley, 13%; oats, 13%; sorghum, 13%; and rice, 12–13%. Like all general rules, this one does not always hold;

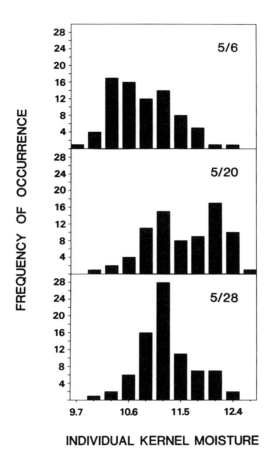

Fig. 2. Distribution of the moisture content of individual wheat kernels from the same sample stored at ambient conditions and sampled on the indicated dates.

the maximum moisture will vary depending upon the temperature, uniformity of moisture in the mass, and other factors.

Drying of Cereals

Drying has been used since early civilization to preserve food. Cereal grains are dried for the same purpose; however, because grain is usually harvested in a relatively dry state, we do not think of it in the same terms. In years past, grain crops such as corn or sorghum were allowed to stand in the field until dry. Crops such as wheat, oats, and rye were cut, bundled, and shocked, then allowed to stand in the field until dry or until a threshing crew became available. The advent of the combine harvester, or the picker-sheller in the case of corn, largely did away with that practice.

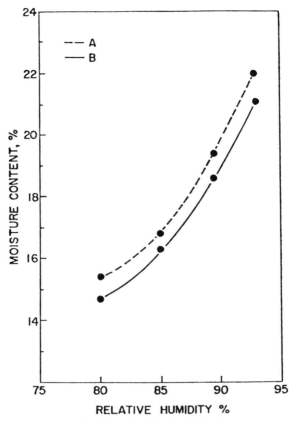

Fig. 3. Moisture content as a function of relative humidity for two corn samples.

Because we cannot predict the weather and because bad weather can destroy the crop left in the field, it is usually wise to harvest the crop as early as possible. In "normal" years, most of the small cereal grains dry rapidly in the field and thus are of a safer storage moisture at the time they are combined. In wet harvest years, however, a considerable amount of small grains must be dried. Even in normal years, most of the corn and rice harvested is dried.

Two major types of drying, low- and high-temperature, are in current use. Low-temperature drying uses air with no heating above ambient conditions. Air, with the heat contained therein, is forced through the grain mass. The system has several obvious advantages. It is relatively energy efficient, requiring only the energy to force the air through the grain. It may also have the advantage of cooling the grain, which is also important for safe storage. Another major advantage of ambient drying is that the grain is not damaged by high heat, as is too often

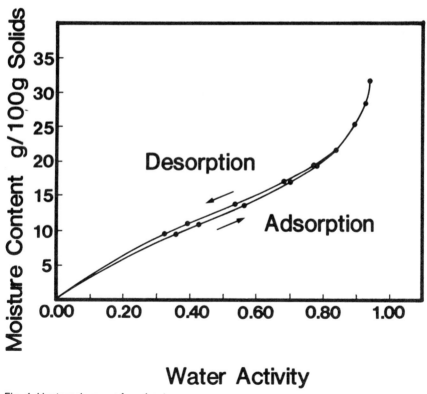

Fig. 4. Hysteresis curve for wheat.

the case with high-temperature drying. The major disadvantage is the relatively long time required to reduce the moisture content significantly. Another obvious disadvantage is that, if the air used has a high relative humidity, that could conceivably increase the moisture content of the grain. Because of its economic advantages, ambient air drying is used wherever possible (Fig. 5). This is usually when the temperature is low enough that moist grain can be safely stored for a period long enough for the ambient drying to be effective.

The drying process can be speeded up by heating the air to higher temperatures. This increases the capacity of the air to hold water and speeds the removal of the water from the grain. The major advantage is, of course, the savings of time.

The major disadvantages are the cost of energy for heating the air and the heat damage the grain may suffer. Such damage may include: stress cracks, increased brittleness and susceptibility to breakage, bulk density changes, discoloration, and loss of germination ability. Less obvious changes also occur and can be seen only when the grain undergoes processing. Wet millers do not like artificially dried corn, as the drying materially affects the ease of separating the starch and protein. Dry millers of corn report that the yield of large grits (their

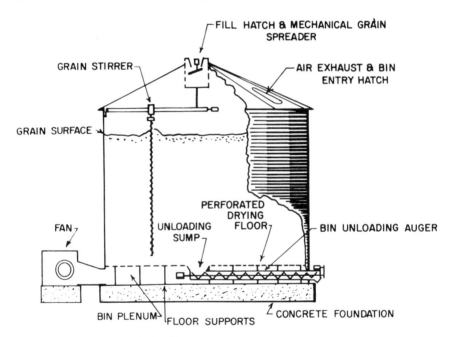

Fig. 5. Low-temperature drying bin.

most valuable product) is materially lower from artificially dried corn. Wheat appears to be very sensitive to heat treatment (Fig. 6) and therefore is very hard to dry economically.

Clearly, the economic pressure is to dry the grain to a safe storage moisture as rapidly as possible, i.e., with more bushels per hour. Corn

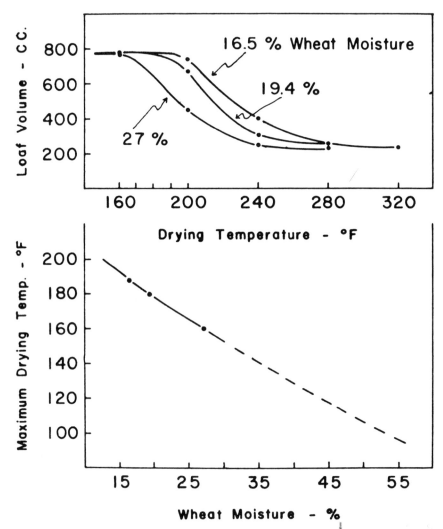

Fig. 6. Interrelations between loaf volume, drying temperature, and wheat moisture content. Maximum drying temperature is the highest that can be employed within a moisture level without impairing loaf volume and crumb grain.

is the grain most commonly dried. Most corn is used as a feed, and because high temperature apparently does not change its feed value, there is little concern about damaging the corn with heat.

Low-temperature drying often is done with a full-bin system. A bin of grain is dried by forcing air through the grain mass. An important consideration in this type of drying is the grain's resistance to airflow. Many factors affect this resistance. Some of these are the size, shape, and surface characteristics of the grain; the size distribution of the grain (which depends, in part, on the number of broken kernels); and the depth of the grain bed. The pressure drop per unit depth of grain is given by the equation:

$$P = cQ^d,$$

where P = pressure drop per unit depth of grain, Q = mass airflow rate per unit area, and c and d are grain-dependent constants. In general, this equation tells us that if airflow is doubled, the pressure increases about three times, and if the grain depth is doubled, the power needed to run the fan increases about 11 times.

The minimum heat required to dry grain is controlled by the amount of water to be removed and the latent heat of vaporization of that water. Other factors that are important are the amount of air used, its initial temperature, its heat capacity, the temperature of the incoming grain, its heat capacity, and the temperature of the outgoing grain.

In drying, the loss of moisture (vaporization) occurs at the surface of the grain. If sufficient energy were supplied to vaporize the moisture inside the kernel, the grain would pop or puff, a clearly undesirable condition. After a short time, the moisture at the surface of the grain has decreased to a low value and the drying rate falls. The rate of drying then depends on the rate of diffusion of moisture to the surface. As drying continues, the distance the moisture must diffuse is greater and the rate of drying is slower.

A system to take advantage of the way grain dries is called *dryeration*. It is actually a combination of heated air drying and aeration cooling. The grain is heated during a high-temperature drying phase, held at elevated temperatures to *temper* (allow the moisture to equilibrate in the kernels), and then cooled and dried with unheated air. A schematic of the process is shown in Fig. 7.

Rice presents a number of problems related to drying. Most importantly, milled rice must be intact if it is to command a premium price. Rice, like most cereals, has a propensity to crack, *fissure*, or check if dried too rapidly. To minimize cracking and fissuring, rice is dried with air heated to less than 65°C. Generally, the grain is dried for only 15–30 min, which may remove 2–3% moisture. The rice is then allowed to temper for 12–24 hr to equilibrate the moisture within the kernel and then is passed through another cycle. The process is continued until the grain reaches the safe storage moisture content of 12.5% or less. Although the process is expensive, as it requires multiple handlings of the grain, it is effective against cracking and checking.

On-farm drying of rice is usually accomplished by using bin-drying techniques with relatively low-temperature air (<10°C). Use of such temperatures and relatively high rates of airflow is effective in reducing the moisture while not fissuring the grain.

Aeration

In years past, grain stored for extended periods of time had to be turned occasionally. Turning, or simply moving it from one bin to another, was to help control grain temperature and eliminate hot spots. In recent years, turning (which is expensive) has been replaced by aeration. This movement of a small amount of air through the mass was found to maintain the temperature satisfactorily. The amount of air is, in general, too low a volume to reduce the moisture content of the grain mass to any extent.

Grains, like most other foods, store better at lower temperatures. Temperatures below 17°C generally prevent insects that attack grains from rapidly increasing in numbers. Microbiological attack is also temperature-dependent. Thus, lowering the temperature of grain that is to be stored is always an advantage.

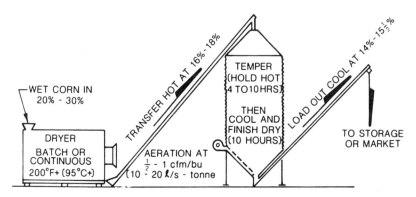

Fig. 7. Schematic of the dryeration process. cfm = cubic feet per minute.

When a bulk of grain is stored in an area where seasonal temperature changes are large, the temperature changes lead to nonuniform temperatures in the grain bulk. Air convection currents are set up that lead to *moisture migration*, resulting in moisture accumulation at particular points in the mass (Figs. 8 and 9). Aeration corrects this condition, as the movement of air through the grain mass makes the temperature more uniform and decreases the moisture accumulation.

Functional Changes and Indices of Deterioration

As long as it retains its viability, grain respires and is thus "alive." Grain stored under reasonable conditions will slowly lose weight

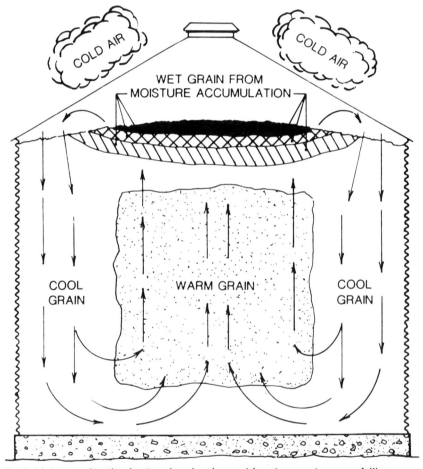

Fig. 8. Moisture migration in stored grain when outdoor temperatures are falling.

because of its respiration. In general, the weight loss is very slow if the grain is stored properly. A much higher proportion of the grain lost in postharvest storage is because of infestation by insects, microorganisms, rodents, or birds. Postharvest losses can be quite large and have been estimated as high as 50% in some countries. Losses in the United States and Canada are, of course, much lower.

Respiration of grain is somewhat difficult to measure because of the difficulty in distinguishing between the respiration of the grain itself and that of the microorganisms that are always associated with the grain. Probably the best way to measure respiration is to measure the CO_2 produced or the oxygen consumed or both. The ratio of moles

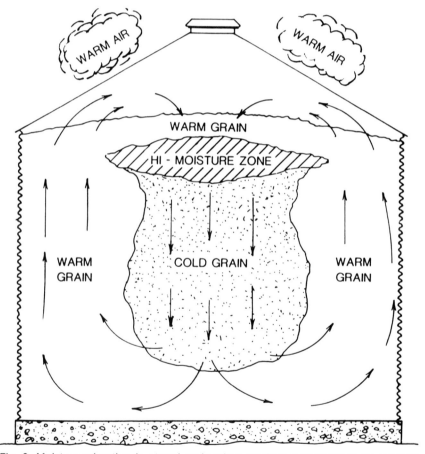

Fig. 9. Moisture migration in stored grain when outdoor temperatures are warmer than the grain.

of carbon dioxide produced to the moles of oxygen consumed is called the *respiratory quotient*. The value of the respiratory quotient varies with the material used as the substrate. It equals 1.0 for carbohydrates and 0.7 for fats.

Respiration can be measured either in a static mode or with gas being forced continuously through the grain mass. Both systems have advantages but, in general, more information is obtained from continuous measurements. Respiration is greatly affected by the moisture content of the grain. Generally, when the interseed relative humidity approaches 75%, the respiration rate increases rapidly. Mold spores are known to germinate at about 75% rh; bacterial growth generally does not start until the relative humidity approaches 90%.

Respiration is also affected strongly by temperature; higher temperature accelerates respiration until thermal inactivation of enzymes starts limiting it. The amount of aeration affects respiration of both the grain and its associated microorganisms, as respiration consumes oxygen and produces carbon dioxide. With low aeration rates, respiration tends to be limited by the oxygen supply. Respiratory activity tends to be higher in grain having a higher percentage of damaged kernels, even with other factors (moisture, temperature, oxygen supply) held constant.

Wheat that has suffered storage deterioration often is called *sick wheat* by the grain trade. The condition is manifested by kernels in which the germ is dead and has turned dark. Such wheat usually has mold growth, but it has been shown that sick wheat can be produced under anaerobic conditions with no mold growth. The darkening of the germ appears to be caused by a Maillard type of browning reaction localized in the germ. As a result of the browning, the germ becomes fluorescent. Although darkening of the germ can occur with no fungal attack, in commercial samples the two are always found together, and the major cause of the loss of germinability is thought to be the fungal attack.

Several tests have been used to measure the condition of grain and thereby predict its future storage behavior. These tests include obvious physical changes in the grain, for example, grain losing its natural luster and becoming dull in appearance. Also easily detected are changes in odor, such as a sour or musty smell. Sick wheat can be identified by the darkened germ (Fig. 10).

One of the first signs of deterioration is a loss of viability. Thus, a germination test can be quite useful, particularly for grain that is to be used for seed or for malting. The value of such tests for predicting changes in grain for such uses as feeding, breadmaking, or breakfast cereals is less clear. A biochemical test for seed viability is based on the reduction, by germ dehydrogenase, of 2,3,5-triphenyltetrazolium chloride. The test is generally called the *tetrazolium test*; a positive result is the appearance of red coloration in the germ.

A number of attempts have been made to develop a reliable and convenient biochemical test to determine storability of grain. These include such assays as fat acidity and glutamic acid decarboxylase activity. Both of these tests are useful, if the users understand what is being measured and correctly interpret the results.

Microflora and Mycotoxins

Cereals are hosts to a large number of different types of microflora (Fig. 11). These include types that invade the seed as well as those that are surface contaminants. The most important of the microflora as far as grain storage is concerned are the fungi, which grow at a much lower interseed relative humidity than other microflora. If the system gets out of control and the moisture content in local regions of the grain increases, other types of organisms can become important, but under good storage conditions this does not occur.

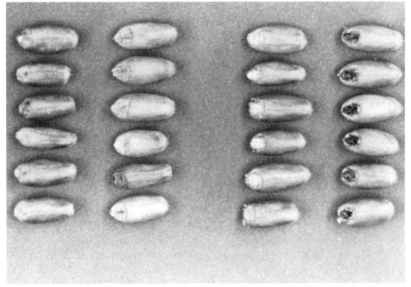

Fig. 10. Intact sound wheat (left row), intact sound wheat with germ exposed (second row from left), intact sick wheat (third row from left), and sick wheat with germ exposed (right row).

118 / CHAPTER FIVE

The storage fungi are always present. Because their characteristics are well known, the conditions necessary to stop their growth have been established. Prevention of damage is then simply a matter of keeping the conditions under proper control. Storage losses to microorganisms can be controlled under almost any environment.

Only a few species of fungi attack stored grain. These are primarily species of *Aspergillus* that are adapted to living on low-moisture grain. Certain species of *Penicillium* also grow on grain at an only slightly higher moisture content. Other species grow only on grain containing relatively high moisture. The high moisture may occur in pockets around areas of fungal growth.

Certain fungi are capable of producing toxic compounds. Some of these are exceedingly toxic when consumed or, in certain cases, when they come in contact with the skin. Some *Fusarium* toxins have been shown to kill mice or rabbits within 24 hr after being applied to the skin. In 1960, the death of a large number of turkeys in England led to the identification of an *aflatoxin* produced by *A. flavus* on peanut meal. Subsequently, four aflatoxins have been identified.

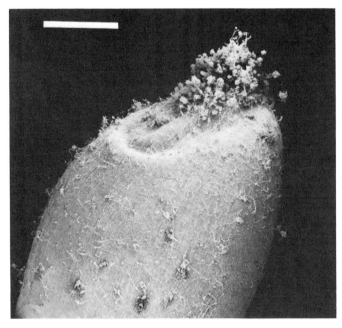

Fig. 11. Scanning electron micrograph of a wheat kernel, showing attachment by microflora.

The presence of mycotoxins in grain presents problems in grain usage. The United States, along with many other countries, has established tolerances for aflatoxins in foods and feeds. One problem is regulation of the permissible amount. A second problem is the relatively large number of lawsuits that have resulted from "toxins" in feeds. These cases are difficult to resolve because low levels of toxins may not kill the animals but may make them perform poorly over an extended period of time (i.e., have poor weight gain). The feed suspected of being toxic may have been consumed long before. Even if the feed were available and were shown to contain organisms that can produce the toxin, this is not evidence that the toxins were indeed present. Both the organism and the right conditions are necessary for toxin production. Much of the mycotoxin produced is the result of field infestation of the grain, although mycotoxin production can also possibly occur from storage organisms.

Insects

Insects are a major problem for the storage of grains and seeds. Not only do insects consume part of the grain, but they also contaminate the grain and thereby constitute a major sanitation problem. The U.S. Department of Agriculture has estimated that storage losses due to insects exceed 470 million dollars per year. Most of those losses could be avoided if the available information on storage were utilized.

Insects that can live on grain can be divided into those that develop inside the kernels and those that live outside the kernels. Those that develop inside the kernels are responsible for the *hidden infestation* found in stored grain (Fig. 12). Five species (granary weevils, rice weevils, maize weevils, lesser grain borers, and Angoumois grain moths) are responsible for that hidden damage. Kernels provide little evidence of the presence of insects inside the kernels until they emerge as beetles or moths. Weevils deposit their eggs inside the kernels. Lesser grain borers and Angoumois grain moths lay their eggs outside the kernels, but the newly hatched larvae tunnel into the kernels to feed and develop.

Insects that develop outside the kernels often feed on broken kernels, grain dust, etc. Important species of this group include confused and red flour beetles, saw-toothed grain beetles, cadelles, khapra beetles, and Indian-meal moths.

Most grain-damaging insects are of subtropical origin and do not hibernate. Thus, their damage can be limited by low temperatures. Not only is low temperature lethal in itself, but it also makes insects inactive so that they do not feed. Generally, temperatures below about

10°C limit the growth and development of most grain-damaging insects. Thus, in the northern part of the United States, cooling the grain to ambient temperatures during the fall and winter is an effective deterrent to grain infestation.

Moisture is another important factor in controlling grain infestation. Insects that live on stored grain depend upon the moisture in the grain for their water supply. Generally, moisture contents of 9% or lower restrict infestation. Interestingly, higher moisture contents also limit infestation because fungi grow rapidly and destroy the insects.

Even though the grain is stored at a relatively safe storage moisture of 11–14%, in the presence of insects, the grain often "heats." The heat is caused by the metabolic heat of the insects. Because of the increased temperature, moisture migration occurs and results in increased moisture in pockets of grain. This leads to microorganisms growing, and the system rapidly gets out of control, with large amounts of grain being damaged.

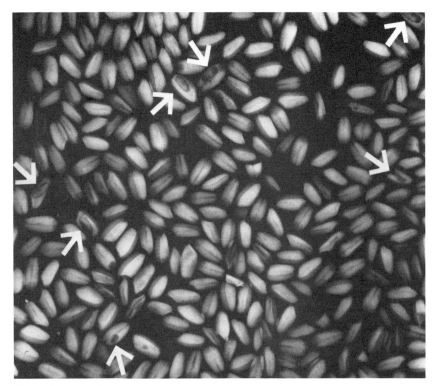

Fig. 12. X-ray of wheat, showing internal infestation (arrows).

Rodents

Next to human beings, rats and mice are probably the most damaging animals known. They destroy (consume or contaminate or both) millions of tons of food each year. Throughout history, rats, mice, and humans have coexisted in an uneasy condition. A popular rule of thumb, which may or may not be true, is that in any given area, the populations of rats and and humans are equal. Estimates have put the rat population of the United States at over 100 million. In one year, a rat may eat as much as 27 lb of food, and this does not account for the food that is made unusable by contamination.

Both rats and mice have a tremendous ability to repopulate. A single pair of Norway rats and their progeny can produce more than 1,500 rats in a year. Thus, trying to kill a rat population, if other factors stay favorable for them, is clearly impossible. Given food, water, and a place to nest, rats and mice will maintain their population.

Killing of rodents, whether by baits, traps, or otherwise, is effective only over short time spans. These methods are helpful in reducing populations or eliminating small populations. The answer to rodent contol is rodent-proofing of buildings and good sanitation, which make the rodents want to find a better place to live.

REVIEW QUESTIONS

1. What is the angle of repose for most cereal grains?
2. What is meant by the term "spoutline"?
3. Explain why moisture is important in grain storage.
4. Why would a grain mass not come to an equilibrium moisture content?
5. Why is an average moisture content of little value to someone wanting to store grain?
6. What is equilibrium moisture content?
7. What is hysteresis?
8. What cereals are usually artificially dried after harvest?
9. What are the advantages and disadvantages of low-temperature drying?

10. What are the advantages and disadvantages of high-temperature drying?
11. How must rice be dried to reduce the tendency for it to crack or fissure?
12. What is the respiratory quotient?
13. What is sick wheat?
14. What is the tetrazolium test?
15. What problems are associated with mycotoxins?
16. What is meant by hidden infestation of grain by insects?
17. What insects that develop outside the kernel are important in causing storage damage?
18. At what temperature do most storage insects become inactive?
19. What is meant by "heating" of grain?
20. What is an effective way of keeping rodents out of grain?

SUGGESTED READING

The book *Storage of Cereal Grains and Their Products*, 4th edition, 1992, edited by D. B. Sauer and published by the American Association of Cereal Chemists, St. Paul, MN, is an outstanding resource in the area of grain storage. Below is a list of its chapters.
Hoseney, R. C., and Faubion, J. M. Physical properties of cereal grains.
Christensen, C. M., Miller, B. S., and Johnston, J. A. Moisture and its measurement.
Pomeranz, Y. Biochemical, functional, and nutritive changes during storage.
Reed, C. Development of storage techniques: A historical perspective.
Bailey, J. E. Whole grain storage.
Brook, R. C. Drying cereal grains.
Foster, G. H., and Tuite, J. Aeration and stored grain management.
Bell, C. H., and Armitage, D. M. Alternative storage practices.
Sauer, D. B., Meronuck, R. A., and Christensen, C. M. Microflora.
Wilson, D. M., and Abramson, D. Mycotoxins.
Harris, K. L., and Baur, F. J. Rodents.
Pedersen, J. R. Insects: Identification, damage, and detection.
Harein, P. K., and Davis, R. Control of stored-grain insects.
Hagstrum, D. W., and Flinn, P. W. Integrated pest management of stored-grain insects.
Manis, J. M. Sampling, inspecting, and grading.
Thompson, S., and Garcia, P. The economics of grain storage.

Other recommended reading:

BAKKER-ARKEMA, F. W., BROOKS R. C., and LEREW, L. E. 1978. Cereal grain drying. Pages 1-90 in: Advances in Cereal Science and Technology, Vol. 2. Y. Pomeranz, ed. Am. Assoc. Cereal Chem., St. Paul, MN.

BAUR, F. J., and JACKSON, W. B., eds. 1982. Bird Control in Food Plants—It's a Flying Shame. Am. Assoc. Cereal Chem., St. Paul, MN.

BULLA, L. A., Jr., KRAMER, K. J., and SPEIRS, R. D. 1978. Insects and microorganisms in stored grain and their control. Pages 91-133 in: Advances in Cereal Science and Technology, Vol. 2. Y. Pomeranz, ed. Am. Assoc. Cereal Chem., St. Paul, MN.

FAN, L. T., LAI, F. S. and WANG, R. H. 1976. Cereal grain handling system. Pages 49-118 in: Advances in Cereal Science and Technology, Vol. 1. Y. Pomeranz, ed. Am. Assoc. Cereal Chem., St. Paul, MN.

LAI, F. S., POMERANZ, Y., MILLER, B. S., MARTIN, C. R., ALDIS, D. F., and CHANG, C. S. 1981. Status of research on grain dust. Pages 237-325 in: Advances in Cereal Science and Technology, Vol. 4. Y. Pomeranz, ed. Am. Assoc. Cereal Chem., St. Paul, MN.

MIROCHA, C. J., PATHRE, S. V., and CHRISTENSEN, C. M. 1976. Mycotoxins. Pages 159-225 in: Advances in Cereal Science and Technology, Vol. 3. Y. Pomeranz, ed. Am. Assoc. Cereal Chem., St. Paul, MN.

Dry Milling of Cereals

CHAPTER 6

Milling is an ancient art. In simple terms, its objective is to make cereals more palatable and thus more desirable as food. Milling generally involves removal of the material the miller calls bran, i.e., the pericarp, the seed coat, the nucellar epidermis, and the aleurone layer. In addition, the germ is usually removed because it is relatively high in oil, which makes the product become rancid faster, thereby decreasing its palatability. The bran and germ are relatively rich in protein, B vitamins, minerals, and fat, and the milled product is lower in these entities than was the original grain. Thus, as a result of milling, the palatability is increased but the nutritional value of the product is decreased. What we call dry milling is the attempt to separate the anatomical parts of the grain as cleanly as possible.

In addition to making the product more palatable and increasing its ability to store longer, milling often involves some type of constraint with regard to particle size. For example, the endosperm of rice or barley must remain in one large piece; from wheat and rye, a fine flour is necessary; a large grit is desirable from corn. Thus, the miller uses a number of procedures to produce the desired product.

Cereals are also wet-milled. Generally, wet milling also attempts to make a clean separation of bran and/or germ from the endosperm, and, in addition, to separate the endosperm into its chemical components of starch and protein.

The Milling Process

GRAIN CLEANING

Grain that is dirty, infested, or "out of condition" because of poor storage conditions should not be used for milling. The cleaning described here is that applied to normal, clean grain. Although large variations are found in the type and number of cleaning steps that different mills employ (Figs. 1 and 2), certain basic steps are necessary.

Early in the system, a *magnetic separator* removes any tramp metal. Actually, a number of magnetic separators are needed, not only to remove metal that would be undesirable in the product, but also to protect machines from damage and reduce sparks that might trigger an explosion. Also early in the cleaning system, a *milling separator*

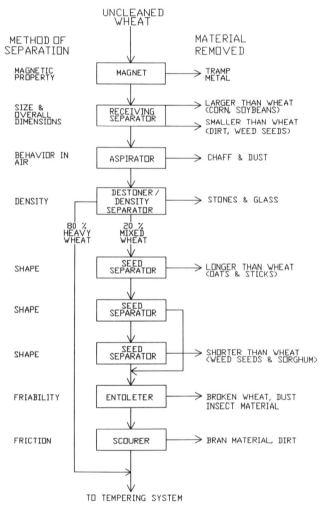

Fig. 1. Simplified flow chart of a wheat cleaning operation.

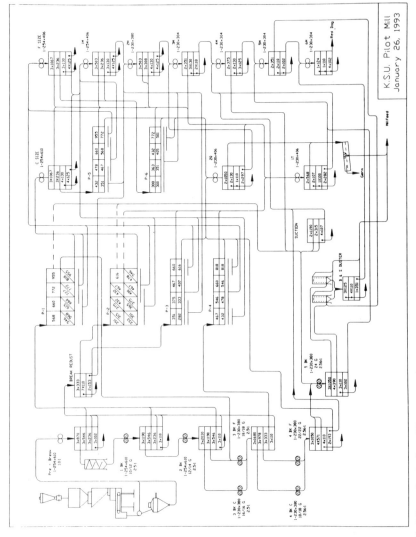

Fig. 2. Simplified flow chart of a milling operation.

removes sticks, stones, and other foreign material that is either larger or smaller than the grain being cleaned. Basically, this machine removes chaff, small pieces of straw, etc., by aspiration, that is, air is pulled through the grain as it is fed into the machine. The grain is then fed onto a sieve larger than the desired grain. This removes large material, including larger grains. The next sieve retains the desired grain but removes smaller grains, by allowing them to pass through. The sieves are reciprocating, i.e., they move back and forth.

In addition to the aspiration found in other machines, specific machines that use airflow to separate light or heavy material from the desired grain may be installed. To separate grains of about the same density from the desired grain, *disk separators* are used (Fig. 3). The separation is based on the length of the kernels. The machine consists of a series of disks mounted a few inches apart on a horizontal central shaft. Both sides of the disk are pocketed. The pockets are designed to pick up grain of a certain length, lift it out of the grain mass, and deposit it into another chamber. The grain enters at one end of the chamber and is conveyed to the other end by the inner spokes of the disk. Grain rejected by the disks then leaves at the end opposite the one from which it entered. Disk separators are generally used in tandem; the first rejects material larger than the desired grain and lifts out the desired grain plus smaller seeds. The second disk rejects the desired grain and lifts out the smaller seeds.

Another important piece of cleaning equipment is the *scourer*. The design varies, but the basic idea is to scour or rub the grain against

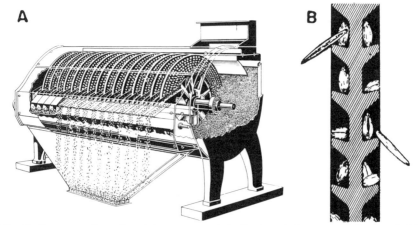

Fig. 3. Cut-away view of a disk separator (A). Grains of the correct size are held in the pockets (B), and large grains are rejected.

itself, a perforated metal screen, or an emery surface to remove adhering dirt. The adhering material loosened by this treatment is removed by aspiration. This machine is useful to remove not only mud but also the smut or rust that result from common wheat diseases.

Small stones similar to the grain in size are difficult to remove. *Dry stoners* of two types are used for this purpose. One is basically an aspirator with sufficient airflow to lift grain. The heavier stones drop out as the grain is lifted. The second type is referred to as a *gravity table* (Fig. 4). The machine separates grains on the basis of differences in their specific gravity, although resilience is also a factor. It consists of a deck supported on rocking legs and oscillated by an eccentric. When the inclination and oscillatory speed of the deck are adjusted correctly, the denser and less resilient material moves down the slope, while the lighter and more resilient material moves up the deck. At one time, washers were used to remove stones as well as loose dirt from the grain; however, they are not used to any extent in the United States today. Several problems have led to their abandonment. The major ones were the control of the amount of water added to the grain as a result of washing, the retention of soluble dirt in the grain, and the difficulty of disposing of the dirty water.

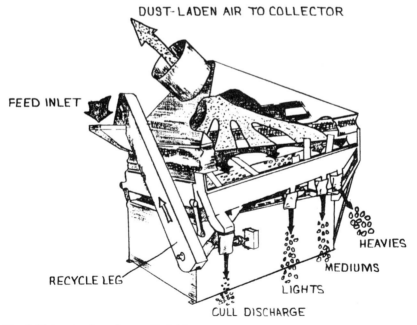

Fig. 4. Cut-away view of a gravity table.

TEMPERING OR CONDITIONING

Tempering consists of adding water to dry grain and allowing the grain to rest for a period of time before it is milled. The purpose of tempering is twofold: 1) to toughen the bran and thus make it resist being broken into small pieces during milling and 2) to soften or "mellow" the endosperm and make it easier to grind. The amount of water added to wheat varies, depending upon its moisture content and the hardness of the grain. Soft wheat is usually tempered to 15.0–15.5% moisture. Hard winter or spring wheats are tempered to 16.5% and durum wheat to even higher moisture levels. The time given for the water to penetrate the grain also varies with the grain hardness. Soft wheat requires a much shorter temper time than does hard wheat.

The penetration of water into wheat has been the subject of much conjecture and study over the years. Recent work using autoradiography to monitor the penetration of tritiated water into the kernel (Fig. 5) has greatly extended our understanding of the phenomenon. In general, such studies show that immediately after wetting, the water is concentrated in the bran and germ. The water causes the cross and tube cells to open up, and the small capillaries hold the water very strongly (Fig. 6). With time, water penetrates into the dorsal region of the grain and finally into the crease area. Diffusion from the bran occurs in all areas of the grain. The rate of water uptake varies for

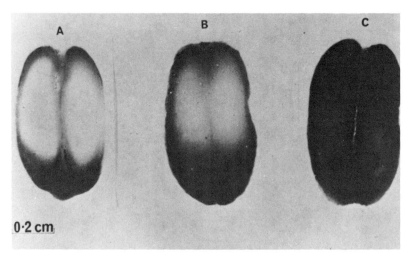

Fig. 5. Autoradiograph of wheat kernels that have taken up tritiated water for 3 hr (A), 6 hr (B), and 24 hr (C).

different cultivars, but the mode of uptake is essentially the same.

No physical barriers to water penetration appear to exist in the bran layers. The outer layers of endosperm, particularly those cells just below the aleurone, appear to be the rate-controlling area for water uptake. Compact endosperm (with no air spaces) appears to slow water uptake. Protein content and hardness of the grain appear to have minor effects on water uptake. Low initial moisture content results in a slower uptake of water. The time required to reach an even distribution of moisture in grain varies from 6 hr for a soft, opaque kernel with low protein to over 24 hr for a hard, vitreous, high-protein kernel. Those times agree well with temper times commonly used in milling.

Bran with high moisture content is tough and tends to stay in larger pieces during milling. This greatly aids in its removal. The effect of water on the endosperm is to make it softer and thus easier to mill. The water presumably breaks, or weakens, the protein-starch bond that is responsible for grain hardness (see Chapter 1). Hard, vitreous wheat becomes soft and opaque as a result of wetting and drying, either in the field because of rain or in the laboratory (Fig. 7).

The term *conditioning* implies the use of heat in conjunction with water to mellow the endosperm. Because the penetration of water into the kernel is essentially by diffusion, the rate is increased by high temperature. However, wheat gluten can be damaged by heat, particularly when the grain is hydrated. Generally, temperatures greater than 50°C should be avoided. In Europe, higher temperatures

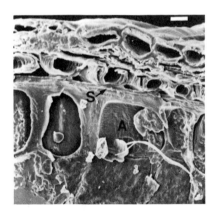 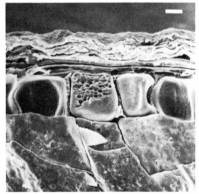

Fig. 6. Outer portion of a wheat kernel: left, wetted and lyophilized kernel; right, dry kernel. The wetting opens the cross and tube cells (T). S, seed coat; A, aleurone cells. Bar is 10 μm.

are used occasionally, not only to speed the uptake of water, but perhaps also to change the gluten's properties. The gluten is less extensible after heat treatment. This change is thought to improve the handling properties of certain European wheat flours.

ROLLER MILLING

The grinding of most cereal grains, particularly those grains having a crease, is done with a roller mill. The roller mill system lends itself to gradual reduction of particle size. The discussion here is for wheat; milling of other grains is discussed later.

The grinding in a roller mill is accomplished on a pair of rolls (Fig. 8) rotating in opposite directions. At the nip (where the two rolls approach each other), the two surfaces are going in the same direction. The rolls are generally run at different speeds, with the fast roll going about two and a half times the speed of the slow roll (the difference is called the *differential*). Thus, in addition to the crushing action as a large particle passes the narrow gap between the two rolls, there is also a shearing action because of the speed differential.

In flour milling, the first objective is to remove the bran and germ from the endosperm. This is accomplished, for the most part, in the *break system* of the mill. After that separation is made, the next objective is to reduce the endosperm to flour fineness. This is accomplished by the *reduction system*.

The break system consists of four or five breaks (sets of rolls). After each grinding, the ground stock is separated on sieves and purifiers

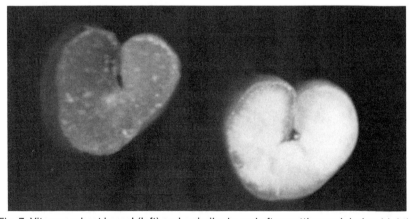

Fig. 7. Vitreous wheat kernel (left) and a similar kernel after wetting and drying (right).

(by air aspiration); the small particles are saved as flour and the large particles are sent on for further grainding. The break rolls are always corrugated. The corrugations are variously shaped cuts that run spirally along the long axis of the roll. The number of corrugations per inch varies from 10 to 12 for first break rolls up to 28–32 at the fourth or fifth break. The action of a set of break rolls can be described as the slow roll holding the material while it is being scraped by the fast roll. In the early breaks, the endosperm is taken off in rather large pieces, and in later breaks, the action is more like a scraping of the bran.

A sifting system (Fig. 9) follows each set of rolls. The stock may be divided on as many as 12 sieves. The large pieces of bran containing considerable amounts of endosperm are sent to the next break; particles

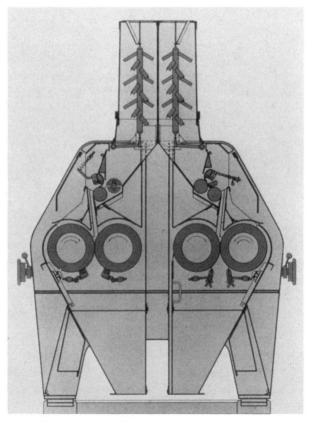

Fig. 8. Roll stand with two sets of rolls.

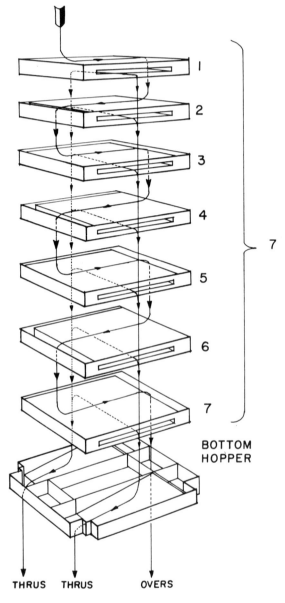

Fig. 9. Illustration of the sieves in a small sieve box, showing the flow of product. 1-7, separations into two parts.

of other sizes may be sent to purifiers and subsequently to reduction rolls. A small amount of flour is produced on each break roll. The medium-sized endosperm particles (middlings) that contain bran particles are sent through a *purifier*. A purifier is essentially an inclined sieve that becomes coarser from the front to the back (head to tail). The sieve is oscillated, and an air current passes upward through the sieve, causing the stock to stratify. Light bran particles are removed by the air; endosperm chunks are graded (separated by size) and sent to the various reduction rolls.

The reduction rolls are generally smooth rather than corrugated. The purpose of the reduction system is to reduce the middlings to flour fineness (so that they can pass a 10XX flour cloth, 136 μm) and remove the last remaining particles of bran and germ. After each grinding pass, the stock is sifted, the flour removed, and the coarser particles sent to the appropriate reduction roll. Purifiers are also used after reduction rolls, mainly to classify the middlings according to size.

The sifting of soft wheat flour is much more difficult than the sifting of hard wheat flour. This statement appears to be illogical. Soft wheat produces a flour with a much smaller particle size; therefore, one would assume that it should sift more easily. This clearly is not the case. Smaller flour particles tend to interact with each other and form aggregates that will not pass a flour cloth (a sieve of silk or other fiber). The interaction forces are liquid bridges, both water and fat on the surface of the flour particles, and frictional forces created by the rougher surface of the flour particles. Soft wheat flour has rougher surfaces than does hard wheat flour, resulting in poorer sifting properties.

Products and Yield

FLOURS

A roller milling operation produces many flours, one from each grinding step. A composite of all the flours produced on the mill is a *straight-grade flour* (Fig. 10). Depending upon the details of the process, it represents about 72% of the total products. The flour produced at the head (first grinding steps) of the reduction system is a *patent flour*. A *short patent* is the flour with the lowest ash and represents about 45% of the total products. This flour comes mainly from the center of the wheat kernel. A *long patent* may be up to 65% of the total products. A *cut-off flour* is the 20% between the short and long patents. The 7% between the long patent and the straight-grade flour is a low-grade flour (clear flour). This flour, which comes from the tail end (last grinding steps) of the breaks and reduction system, has high ash and dark color. Another product from the mill is *millfeed*, (the *bran* and *shorts*); bran represents about 11% of total products and shorts about 15%. Germ can be an additional product, usually recovered at about 0.5%.

Calculation of yield would appear to be simple. Starting with 100 lb of wheat and obtaining 70 lb of flour gives a 70% yield, or extraction—correct? The answer is maybe. Although the above calculation can be used after one has decided whether to use dirty wheat or cleaned wheat as a basis, the more common expression of yield used by millers is based on the flour yield as a percent of total product, not as a percent of the starting material. The percentage of total product in any desired fraction is called the *extraction*. In other words, a straight-grade flour that is 72% of the total product is a 72% extraction flour.

ASH

Typical wheat has an ash (mineral) content of about 1.5%. However, the ash is not distributed evenly in the grain. The inner endosperm is relatively low in ash (~0.3%), whereas the bran is higher in ash and may contain as much as 6%. Thus, ash is a convenient assay for the presence of bran in flour, as its determination is relatively simple and reproducible. However, wheats vary in the amount of ash

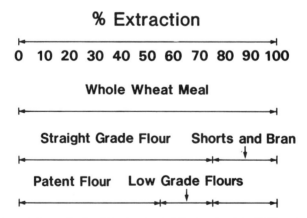

Fig. 10. Grades of flour obtained from wheat. Whole wheat is 100% extraction (line 2) but can be divided into straight-grade flour plus shorts and bran (line 3). The straight-grade flour can be further divided into patent and low-grade flours (line 4).

natively found in the endosperm, and thus small variations in ash content do not necessarily imply the presence of different amounts of bran in flours.

In general, the ash itself does not affect flour properties, and thus it can be argued that the ash content alone has no meaning. Recently, great interest has been shown, particularly in Europe, in using color to replace the ash test. Proponents argue that color has more meaning. Perhaps, but it is still true that high ash is an indication of bran in the flour. It is also true that the bran contains compounds that generally are detrimental to baking performance.

CHANGE IN COMPOSITION AS A RESULT OF MILLING

The bran and germ have chemical compositions that are quite different from that of the endosperm and therefore of flour (Table I). Thus, milling results in changes in composition (Table II). All the compounds reported in Table II except starch and chlorine decrease in concentration as they go from wheat to flour. The changes in the concentration of certain vitamins and minerals provide the rationale for enriching flour and other cereal products. Curves of ash or protein extraction (Fig. 11) are useful in determining the maximum yield one can obtain and still meet ash and protein specifications.

Processing of Flour

AIR CLASSIFICATION AND FINE GRINDING

Most flour is sieved through a 10XX flour cloth. The openings on that sieve are about 136 μm. Thus, flour is composed of a range of particles, from very small (<1 μm) up to about 200 μm (particles larger

TABLE I
Typical Analysis of Mill Fractions[a]

	Wheat	Patent Flour	Germ	Shorts	Bran
Protein	12.0	11.0	30.0	16.0	14.5
Ash	1.8	0.40	4.0	4.1	6.0
Fiber	2.5	...	2.0	5.5	10.0
Fat	2.9	0.88	10.0	4.5	3.3

[a]Adapted from E. Ziegler and E. N. Greer. 1971. Pages 115-199 in: Wheat: Chemistry and Technology, 2nd ed. Y. Pomeranz, ed. Am. Assoc. Cereal Chem., St. Paul, MN.

than the opening can pass through a sieve). Viewed through a microscope (see Chapter 1), flour can be seen to be made up primarily of two components, protein and starch. The smaller particles are composed of pieces of endosperm protein or small starch granules with adhering protein. Thus, the small particles tend to be high in protein. As particle size increases, a larger proportion of the particles are free starch granules, and the protein content is lower. Even-larger particles primarily consist of the starch embedded in a protein matrix, just as was the case in the intact endosperm. That fraction has a protein content equal to that of the original flour. Thus, by separating the flour into particle size classes below those obtainable with normal sieves, we can obtain subsamples of the flour with much higher protein (in particles of <17 μm), lower protein (17–35 μm), and equal protein (17–35 μm) in comparison with the protein level in the original flour. Air-classifiers have been developed to make these separations.

Soft wheat flour air-classifies better than hard wheat flour does. Because of its soft texture, soft wheat produces a flour with a much smaller particle size than hard wheat. In hard wheat, the protein and starch tend to remain together in the flour and cannot be separated without additional grinding to reduce the particle size. Pin mills are

TABLE II
Change in Composition from Wheat to Flour

	Wheat	70% Extraction Flour
Ash, %	1.55	0.4
Fiber, %	2.17	Trace
Protein, %	13.9	12.9
Oil, %	2.52	1.17
Starch, %	63.7	70.9
Thiamin, μg/g	3.73	0.70
Riboflavin, μg/g	1.70	0.70
Niacin, μg/g	55.6	8.50
Iron, mg/g	3.08	1.42
Sodium, mg/g	3.2	2.2
Potassium, mg/g	316	83
Calcium, mg/g	27.9	12.9
Magnesium, mg/g	143.0	27.2
Copper, mg/g	0.61	0.18
Zinc, mg/g	3.77	1.17
Total phosphorus, mg/g	350	98
Phytate phosphorus, mg/g	345	30.4
Chlorine, mg/g	39.0	48.4

[a]Adapted from E. Ziegler and E. N. Greer. 1971. Pages 115-199 in: Wheat: Chemistry and Technology, 2nd ed. Y. Pomeranz, ed. Am. Assoc. Cereal Chem., St. Paul, MN.

effective in this reduction of particle size. At high speeds, pin mills produce considerable amounts of fractured starch, which generally is undesirable.

STARCH DAMAGE

During milling, a small but significant number of the starch granules in the flour are damaged. The level of damage varies with the severity of grinding and the hardness of the wheat. A soft wheat can be milled with essentially no starch damage. Even so, because of the analytical methods used to measure starch damage, the value may be 2–3%. The procedure approved by the American Association of Cereal Chemists for measuring starch damage is based upon the fact that damaged starch is susceptible to fungal α-amylase and undamaged starch is not.

There are also different types of starch damage. A starch granule can be broken in two, as illustrated in Fig. 9 of Chapter 1. Although

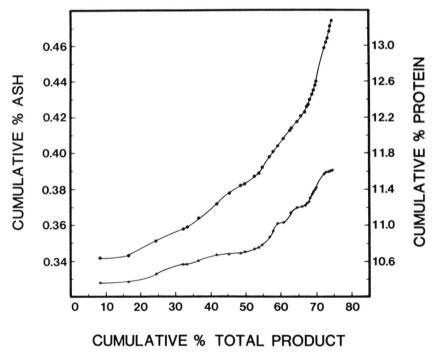

Fig. 11. Curves of ash extraction and protein extraction for a commercial wheat flour mill. Upper curve, ash; lower curve, protein.

the granule is clearly damaged, this type of damage results in starch that is still birefringent, not soluble in water, and not susceptible to enzymes; it doesn't act damaged. The more classic starch damage produced during milling results in granules that have lost birefringence, appear as "ghosts" under the microscope, stain with Congo red, and are susceptible to fungal α-amylase.

People often state that damaged starch is necessary in breadmaking; however, why this would be true is unclear. Perhaps in formulas containing little or no added sugar, damaged starch could be helpful; however, bread formulas used in the United States practically always have sufficient sugar added so that the level of damaged starch is not important from the standpoint of gassing power. Damaged starch increases the water absorption of dough. It also produces weak side walls and a sticky crumb if sufficient amylolytic enzymes are available. Starch damage is a strong negative factor in soft wheat flour used as cookie flour.

FLOUR TREATMENTS

Flour may receive a number of treatments at the mill. Bleaching is very common. For bread and all-purpose flours, the most common bleaching agent is benzoyl peroxide, which is added as a dry powder and whitens flour over a two-day period. It only bleaches flour pigments and has no effect on the baking properties. Cake flour is often treated with chlorine gas, which immediately destroys (bleaches) the pigments and gives a very white flour.

Chlorine is detrimental to bread flours but beneficial to cake flours. High-ratio cakes (those that contain more sugar than flour) cannot be made without treatment with chlorine gas. The pH of flour drops when it is treated with Cl_2. The lower pH is not beneficial in itself but gives a way of monitoring the amount of reaction between the gas and the flour. Thus, with cake flours, Cl_2 is both a bleach and an improving agent.

When flour is stored long enough after milling, it slowly bleaches (supposedly the result of air oxidation). It also undergoes aging or maturing (improvement in baking performance) with long storage times. A number of oxidizing agents are used at the mill as maturing agents. These include azodicarbonamide, acetone peroxide, chlorine dioxide, and potassium bromate. All of these reagents improve the bread-making capacity of flour. They are discussed in more detail in Chapter 12.

Most wheats milled in America are low in α-amylase. To correct this problem, malted barley or malted wheat flour is added, usually at about 0.25% of the flour. The proper treatment level can be determined by the amylograph or falling number tests. Treatment improves loaf volume and reduces the harshness (rough texture) of the crumb.

Since 1940, all "family" flour and baker's bread flour have been required to be enriched with the vitamins thiamin, riboflavin, and niacin and with the mineral iron. The purpose is to replace part of the vitamins and minerals lost as a result of milling.

Self-rising flour is also prepared at some mills. The flour contains added sodium bicarbonate, acid salts (either monocalcium phosphate or sodium acid pyrophosphate or both), and salt (see the discussion of chemical leavening in Chapter 13). The acid added is sufficient to neutralize the sodium bicarbonate. Self-rising flour must contain enough leavening to evolve an amount of carbon dioxide not less than 0.5% of the flour weight. Self-rising flour is used to make American biscuits. All that is needed is to add water or milk to the self-rising flour.

AGGLOMERATION

Flour can be agglomerated by wetting the outside of the flour particles, bringing them in contact with other particles, and drying them. This is accomplished by agitating flour in an atmosphere of water droplets followed by drying in an air stream. The agglomerated flour is dust free, has a controlled bulk density, has good flowability, and disperses in water without lumping.

Milling of Grains Other than Wheat

DRY CORN MILLING

The corn kernel presents a number of problems for the miller. It is large, hard, flat, and, in addition, contains a larger germ than other cereals (~12% of the kernel). The germ is high in fat (34%) and must be removed if the product is to be stored without becoming rancid. In the traditional stone-grinding of cornmeal, the germ is not removed; however, even though the product is generally not shipped very far, it is often rancid. In corn milling, the desired products are low-fat grits rather than corn flour. Thus, the miller wants to remove the hull (that is, the pericarp, seed coat, and aleurone layers) and germ without reducing the endosperm to small particle size. The most effective way to accomplish this is with a *degerminator*, a specialized attrition mill. It is basically two cone-shaped surfaces, one rotating inside the other. They rub the corn to remove the hull and break the germ free. Generally, the corn is tempered to a high moisture content (~21%) before it is degermed. This allows easier separation of the germ from the endosperm.

Entoleters have also been used for degerming. The entoleter is an impact device rather than an attrition mill. The corn enters the machine by falling on a rapidly rotating disk containing pins. It is forcefully thrown against the wall and degermed by the impact.

After the germ and hull are removed, the endosperm is reduced to grits of the desired size by roller milling much in the fashion of wheat milling. It is usually dried to obtain a desired moisture content. The dry corn mill is basically a break system, as flour is not the desired product. The hulls are sold as animal feed, and the germ is processed to recover the valuable oil.

RYE MILLING

Rye milling is similar to wheat roller milling but differs in several aspects. Rye selected for milling usually has less than 8% thin kernels (those passing a 1.6- × 9.5-mm screen). Thin kernels yield a very low amount of white flour. During cleaning, care must be taken to remove ergot (the fruiting body of *Claviceps purpurea*). Rye is more prone to contain *ergot* than other cereals are.

Ergot is a fungus that replaces the kernel in the head. It is roughly the same size and density as a grain kernel and therefore is difficult to remove during cleaning. Ergot contains certain alkaloids that are quite toxic. They can result in death if sufficient quantities are consumed. However, ergot is also a valuable by-product that can be sold to the pharmaceutical industry.

In general, rye kernels are soft, and water penetrates them rapidly. Therefore, rye is generally milled after a short temper of 6 hr to a moisture content of 15.0%. In a rye mill, all the rolls are corrugated, as a smooth roll tends to flake the product rather than crack it. Because of its soft texture, rye flour sieves with difficulty, and thus a large amount of sifter surface is needed. Purifiers are not used in rye mills. In general, two types of rye flour are produced in the United States: a white rye, which makes up about 80% of the total flour produced, and a dark rye that constitutes the other 20%. Other flours are produced by a combination of these two basic flours. Some of the white rye flour is treated with a small amount of chlorine gas as a bleach.

Rye flour doughs do not have the gas-holding ability of wheat flour doughs. Therefore, rye bread made entirely from rye flour is very dense.

Most rye bread produced in the United States contains relatively small amounts of rye flour for flavor but large amounts of wheat flour to produce the light texture. Rye flour contains large amounts of pentosans compared to other cereal flours. The pentosan-starch ratio is generally believed to be the primary factor responsible for the baking quality of rye flours.

DECORTICATION OR ATTRITION MILLING

Decortication is the removal of the "bark," or outer layers, of the grain. For those grains that do not have a crease, this type of dry milling offers an alternative to roller milling. When the desired product is the whole grain with the bran removed, rather than flour, decortication is the only alternative. Examples of this are milled rice and pearled barley. Rice and oat milling are discussed more fully in a separate chapter.

Other grains that are often decorticated are barley, sorghum, and millet. Barley is harvested with the hull on. During pearling, the hull, pericarp, seed coat, and aleurone are removed, leaving the remainder of the endosperm relatively intact. Pearled barley is then reduced to flour fineness by a hammer mill or used as the intact endosperm in soups. The pearling device has an abrasive surface, and the outer layers are essentially sanded off the grain. This is an attrition process; the amount removed from the grain can be controlled by the length of time the grain remains in the machine (see Fig. 3 of Chapter 8).

In areas where they are consumed as food, mainly India and Africa, sorghum and millets are often milled by decortication processes. Decortication appears to be used for two reasons. First, the grains are nearly round and do not have a crease. Second, the decortication process is similar to hand-pounding with a wooden pestle in a mortar, which was the traditional processing of the grains. Most of the users of sorghum and millet use a coarse ground flour rather than the intact endosperm; thus, after decortication, the endosperm is reduced by further pounding or by milling with either a hammer mill or burr mill. One major problem with the decortication process is that the germ is often left with the endosperm. The presence of the high-fat germ often leads to the flour rapidly becoming rancid.

REVIEW QUESTIONS

1. What is the miller trying to accomplish during dry milling?

2. What principles of separation are used in a milling separator, a disk separator, and a scourer?
3. How does a gravity table work?
4. What is tempering and what is its purpose?
5. Does hard or soft wheat require a longer temper time and why?
6. What is the break system of the mill?
7. What is a purifier and how does it work?
8. What is the difference between break rolls and reduction rolls?
9. Why is soft wheat flour much harder to sift than hard wheat flour?
10. What is straight grade flour? a patent flour?
11. What does high ash in a flour indicate?
12. What can be learned from an ash extraction curve?
13. Explain how an air classifier can give flours of different protein contents from a single flour.
14. Explain what damaged starch is and how it is produced during milling.
15. What is benzoyl peroxide and what does it do to flour?
16. What does chlorine gas do to wheat flour and why is it used?
17. List a number of maturing agents that might be added at the mill.
18. Why is malted wheat or barley flour added to bread flours?
19. What is added to flour to make it self-rising?
20. What are the desired products from a dry corn mill?
21. Why are all the rolls in a rye mill corrugated?
22. What is decortication?

SUGGESTED READING

ALEXANDER, R. J. 1987. Corn dry milling: Processes, products, and applications. Pages 351-376 in: Corn: Chemistry and Technology. S. A. Watson and P. E. Ramstad, eds. Am. Assoc. Cereal Chem., St. Paul, MN.

BASS, E. J. 1988. Wheat flour milling. Pages 1-68 in: Wheat: Chemistry and Technology, 3rd ed. Vol. 2. Y. Pomeranz, ed. Am. Assoc. Cereal Chem., St. Paul, MN.

HALVERSON, J. and ZELENY, L. 1988. Chriteria of wheat quality. Pages 15-45 in: Wheat: Chemistry and Technology, 3rd ed. Vol. 1. Y. Pomeranz, ed. Am. Assoc. Cereal Chem., St. Paul, MN.

LARSON, R. 1970. Milling. Pages 1-42 in: Cereal Technology. S. A. Matz, ed. Avi Publishing Co., Westport, CT.

LOCKWOOD, J. F. 1952. Flour Milling, 3rd ed. The Northern Publishing Co., Ltd., Liverpool.

ROZSA, T. A. 1976. Rye milling. Pages 111-125 in: Rye: Production, Chemistry, and Technology. W. Bushuk, ed. Am. Assoc. Cereal Chem., St. Paul, MN.

SCOTT, J. H. 1951. Flour Milling Process, 2nd ed. Chapman and Hall, Ltd., London.

SHAW, M. 1970. Rye milling in the United States. Assoc. Oper. Millers Tech. Bull. 3203-3207.

SHELLENBERGER, J. A. 1980. Advances in milling technology. Pages 227-270 in: Advances in Cereal Science and Technology, Vol. 3. Y. Pomeranz, ed. Am. Assoc. Cereal Chem., St. Paul, MN.

WELLS, G. H. 1979. The dry side of corn milling. Cereal Foods World 24:333, 340-341.

ZIEGLER, E, and GREER, E. N. 1971. Principles of milling. Pages 115-199 in: Wheat: Chemistry and Technology, 2nd ed. Y. Pomeranz, ed. Am. Assoc. Cereal Chem., St. Paul, MN.

# Wet Milling: Production of Starch, Oil, and Protein

CHAPTER 7

Dry milling is primarily concerned with separation of the anatomical parts of the grain. Wet milling strives for the same separations but also goes a step further and separates some of those parts into their chemical constituents. Thus, the primary products are starch, protein, oil, and fiber instead of bran, germ, and endosperm.

Corn

After a cleaning similar to that used in dry milling, the corn is steeped. Steeping involves submerging the corn in water containing 0.1–0.2% sulfur dioxide. The temperature is elevated and controlled at 48–52°C, and the steep time varies from 30 to 50 hr. As a result of steeping, the corn contains about 45% moisture and is softened sufficiently so that the softness can be detected by squeezing. About 6% of the corn kernels become soluble during the steeping process. Commercially, corn is steeped in tanks that may hold 3,000 bu of grain. The steeping system normally uses about 10 tanks, with the corn moving from tank 1 to tank 10 and the steep water moving from tank 10 to tank 1.

Sulfur dioxide, often generated by burning sulfur, is used for two reasons. First, it aids in stopping the growth of putrefactive organisms. Second, the bisulfite ion reacts with disulfide bonds in the matrix proteins of the corn (Fig. 1) and reduces the molecular weight of the proteins, making them more hydrophilic and more soluble. As a result, the release of starch from the protein matrix is much easier and the yield of starch is higher. During steeping, the level of sulfur dioxide in the steep water decreases as more of the bisulfite ion reacts with the protein. Corn that has been dried at excessively high temperatures gives lower amounts of soluble protein when steeped with bisulfite than does unheated corn. Heat-damaged corn also gives greatly reduced yields of starch and therefore is not desired for wet milling.

Although sulfur dioxide slows the growth of some organisms, it does not stop certain lactobacilli. Steeping at 45–55°C favors production of lactic acid organisms. Lower temperatures lead to production of butyric acid. The corn itself appears to be the source of the organisms. The role of the lactic acid produced during the steep is not clear. It appears to have only a minor effect on the softening of the corn kernel. Perhaps its major effect is to lower the pH and stop the growth of other organisms.

After steeping, the steep water contains about 60 g of solubles per liter. Generally, the steep water is concentrated (by reverse osmosis membrane filtration) to about 55% solids and mixed with the hulls (or, more correctly, the bran) and/or the spent germ to be used as animal feed. The solids from the steep also make a good growth medium for the production of desired microorganisms. The dried solids contain about 35% protein nitrogen, 26% lactic acid, 18% ash, and 7% phytic acid. In addition, they contain reasonable levels of B-vitamins.

After steeping, the softened grain is coarsely ground in water by an attrition mill. The object here is to break the grain and free the germ without breaking it into pieces. As a result of steeping, the germ is swollen and rubbery. Two passes through the mill may be needed to free the germ, after which it is separated from the remainder of the kernel with a liquid cyclone separator, or *hydroclone* (Fig. 2). The separation is based on density; the germ has a lower density because of its oil content. The hydroclone works in water on the same principle

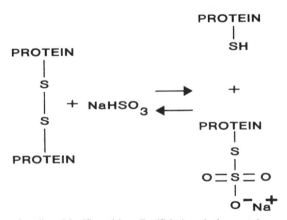

Fig. 1. Reaction of sodium bisulfite with a disulfide bond of a protein.

as a cyclone separator does in air. The recovered germ is washed free of adhering starch and dried. It is then used to produce oil.

After removal of the germ, the remaining material is sieved. The coarser particles, the hulls (bran) and endosperm chunks (mainly from the hard, horny endosperm), are ground again. This grinding is accomplished on stone or steel burr mills or with an impact type of mill. The object is to free the starch, protein, and fiber from each other. The fiber (bran) tends to stay in larger pieces and is removed by screening. Generally, the fiber is given a series of screenings on various sized screens and is washed to remove adhering starch. The finest screen may be 75 μm. After the fiber is washed, it is dewatered (with pressure) and then dried for use in animal feeds.

The remaining stream contains the protein and starch in water. Because the starch is denser than the protein, they can be separated from each other with large continuous centrifuges or with additional hydroclones. The less dense gluten, containing 60–70% protein on a dry basis, is dewatered by centrifuges and then dried. It is a valuable by-product used as an animal feed.

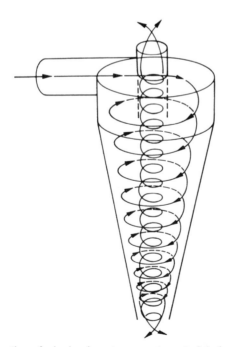

Fig. 2. Operation of a hydroclone to separate material of different densities.

The starch still contains too much protein (~1%) at this point and must be purified by recentrifuging or with hydroclones. The hydroclones used here work on the same principle as those used to separate germ; however, they are much smaller in size, and large numbers of them are used in sequences. Starch coming from them contains less than 0.3% protein and is ready for chemical modification, conversion to syrup, or drying to be sold as starch. Most drying is accomplished with flash dryers. The dewatered starch is injected into a rapidly moving stream of heated air. The granules are dried rapidly and collected with dust cyclones.

A schematic diagram of the corn wet-milling process is given in Fig. 3. Notice that all the water (about 10 gal/bu of corn) enters at the final washing step and works its way back to the steep tanks. The concept of countercurrent flow of water and corn is sometimes difficult to grasp. It is not water flowing one direction in a pipe and corn flowing another. As the starch is washed, the wash water removed from the starch is used in a preceding step of the flow. This process is then repeated until the water is in the steep tanks. There are no effluent streams; the system is bottled up, and water leaves the plant only as water vapor.

Sorghum can be and has been wet milled by essentially the same system as described above for corn. Three major problems with sorghum are that 1) it has a small germ (so no valuable oil is recovered), 2) the presence of pigments (mainly polyphenolics) in the pericarp can give the starch an off color, particularly if it is placed in an alkaline solution, and 3) the bran of sorghum breaks into small pieces that interfere with the separation of the protein and starch.

Wheat

Although the major source of starch in the United States is corn starch, a significant amount of wheat starch is produced. Even though wheat starch does have properties different from those of corn starch, those differences probably do not account for its production. Wheat starch is probably produced more because of the coproduction of the valuable wheat gluten fraction than because of any desire to have the wheat starch.

At the present time, low-grade flour rather than wheat is the starting material for the production of wheat starch and gluten. Two similar processes are used in the United States. A flour-water dough is mixed (a stiff dough in the Martin process and a batter in the batter process). The next step is to wash the gluten from the starch. When hydrated, wheat gluten forms a strongly cohesive mass. By gently kneading the dough in water, one can separate the starch from the gluten. Since the gluten tends to adhere to itself and stay in large pieces, it can be sieved from the starch. The gluten can be enriched to about 75% protein (dry basis) by additional washings with water. The starch, which is purified by the use of large continuous centrifuges, is composed

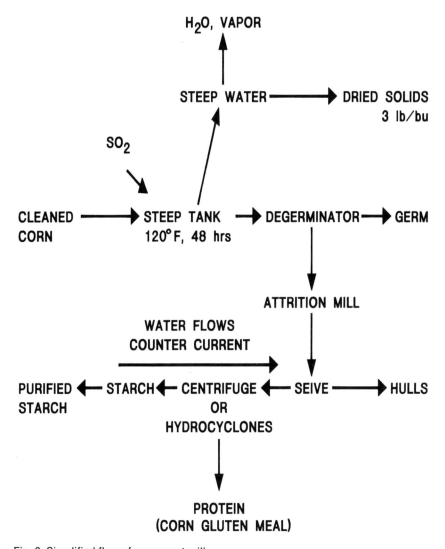

Fig. 3. Simplified flow of a corn wet mill.

of two fractions. The prime starch, called *A starch* in the industry, contains the large, lenticular starch granules and part of the small, spherical starch granules. The second fraction, called *B starch*, is composed of small starch granules, pentosans (cell wall material), and damaged starch granules. One of the disadvantages of starting from a flour to produce gluten and starch is that flour contains starch that was damaged during the dry-milling process. The B starch, which is also called tailings starch or squeegee starch, makes up as much as 20% of the total starch. This fraction is of much lower value than the A starch.

During the processing of flour, as much as 10% of it becomes soluble. Apparently this soluble material is not recovered. Because of the presence of water-soluble pentosans, the fraction cannot be concentrated in a manner similar to that used for the solubles from corn steep, as the viscosity would become very high.

Sulfur dioxide is not used in a wheat starch operation for two reasons. First, it is not needed, as water alone softens the wheat flour particles and allows separation of the protein from the starch. Second, the sulfur dioxide would destroy the character of the *vital wheat gluten* and make it much less valuable. *Vital* is the term used in the industry to identify gluten that has not been damaged and thus has the ability to form a dough that will retain gas.

A major problem in starch-gluten plants is the drying of the products. Wheat starch, particularly the small-granule starch, is like grain dust in that it can form a very explosive mixture. Unless precautions are taken and adhered to, an explosion is likely. Several wheat starch plants have had explosions in the last few years. Wheat gluten is valuable because of its vital character. Heating gluten, particularly after it has been hydrated, will alter the protein, making the gluten no longer vital and lowering its value considerably. Most gluten plants use a type of flash dryer that purports not to damage the gluten. The basic process is to extrude wet gluten into a stream of heated air containing previously dried gluten. The concept is that evaporating water will keep the product cool until it is dry. The theory is good; in practice, most of the vital wheat gluten on the market has lost a significant part of its vital character.

Over the years, numerous attempts have been made to produce gluten and starch from wheat and not from dry-milled flour. Starting with the whole grain has several advantages. For one, the cost of and damage produced by dry milling are avoided. For another, when buying wheat, one can specify the protein content and wheat type, but with low-grade (clears) flours, one takes what is offered. Patents were issued in the 1970s to Pillsbury Co. and to Far-Mar-Co for processes to produce gluten and starch from whole wheat. Both companies had pilot plants or slightly larger plants operating. At the present time, no significant amounts of starch and gluten are produced directly from wheat.

Rice

Rice starch is apparently not produced commercially in the United States. Although it is a starch of commerce in Europe, relatively small amounts are imported into the United States.

As shown previously (Chapter 1), rice starch occurs as compound granules tightly held in a protein matrix. The individual granules are very small (4–8 μm in diameter) and polygonal in shape. The protein-starch association in the rice kernel is very strong and is not weakened when the grain is soaked in water or sodium bisulfite. However, it is weakened by soaking in dilute sodium hydroxide.

The basic process to produce rice starch consists of soaking broken rice (often pulverized first by dry milling) in 0.3% sodium hydroxide for about 12 hr. Most of the rice protein is solubilized by this process. The starch is recovered by centrifugation. After a series of washings with water, the starch is dried. The protein, recovered from the combined steep solution and wash water by neutralization with hydrochloric acid, is used as a supplement for cattle feed.

Production of Oil from Cereals

Cereals vary widely in the amount of oil they contain (Table I). However, in all cases, the amount of oil is too low to justify its recovery. Thus, the production of oil is as a by-product of other manufacturing processes.

TABLE I
Composition of Rice Bran and Polish[a]

Constituent	Bran	Polish
Protein	12.0	12.0
Fat	13.0	16.0
Ash	10.0	8.0
NFE[b]	40.0	56.0
Crude fiber	12.0	7.3
Pentosans	10.0	...

[a] Adapted from D. F. Houston. 1972. Pages 272-300 in: Rice: Chemistry and Technology, 1st ed. D. F. Houston, ed. Am. Assoc. Cereal Chem., St. Paul, MN.
[b] Nitrogen-free extractives.

The amount of oil produced from cereals is small compared to that produced from oilseeds and from animal sources. Some oil is produced from corn, wheat, and rice. Wheat oil is more unsaturated than the other cereal oils. In general, the higher the degree of unsaturation, the more problems encountered in storing the oil. In the United States, wheat oil is sold mainly in health food stores as a nutritional supplement. Rice oil, as one might expect, is found in Asia. The oil is extracted from the bran rather than from the germ as is the case for the other cereals.

The major cereal oil produced is from corn germ. Corn oil generally sells at a higher price than the other oils. The reasons for the premium are that corn oil has a good nutritional image (unsaturated corn oil is good for you) and that it can be refined to a light-colored and light-tasting oil that most people prefer. Thus, most corn oil is consumed as a salad or cooking oil, with smaller amounts used in margarine. Practically none is converted to shortening.

Corn germ from either wet-milling or dry-milling plants is used for oil extraction. Generally, the germ is flaked to give a coarse meal, heated with steam, and run through an expeller unit. In the expeller, the germ meal is subjected to high pressure in a slotted barrel. Most of the oil is pressed out through the slots, and the residue of the germ is discharged at the end of the barrel. In several passes, perhaps up to 95% of the original oil can be expelled. The residue from the expeller, called *foots*, is often solvent-extracted with hexane. The resultant residue contains about 0.5% oil.

After coming from the expeller or after being stripped to remove the solvent from the solvent extract, the crude oil is filtered to remove solid impurities. The filtered oil is treated with a strong alkaline solution, usually at high temperature (93°C) for a short time. Longer treatment must be avoided to reduce saponification. The object of the treatment is to convert the free fatty acids and phospholipids to water-soluble salts. These two lipid classes, which are then removed in the aqueous wash, are called *soap-stock* and are used in soap manufacture. The free fatty acids must be removed because they reduce the storage stability of the oil, the phospholipids because they lower its smoke point. The crude oil generally contains about 0.5% free fatty acids, which is reduced to about 0.01% after the alkali treatment. The crude corn oil also contains about 1.5% phospholipids; this level is reduced to less than 0.05%.

After the alkali treatment, the oil is "bleached" (Fig. 4). Actually, it is not a bleaching at all, but the absorption of pigments on an absorbent. Acid-activated clay (about 2% of the amount of the oil) is vigorously mixed with oil at about 105°C. After a short equilibrium period, the oil is filtered to remove the clay (or oil can be recovered from the clay by extraction). The clay is then discarded.

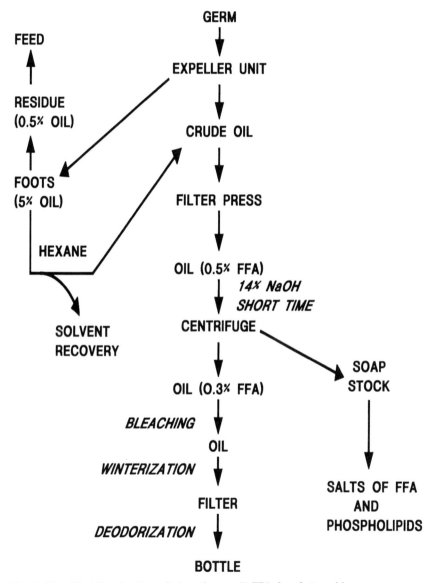

Fig. 4. Simplified flow for the refining of corn oil. FFA, free fatty acid.

The next step in the oil-refining process is *winterization*. Its object is to remove material that becomes insoluble at lower temperatures and produces an oil that appears cloudy. The cloudy appearance is undesirable, particularly in salad oils, which are often refrigerated. In corn oil, the cloudy appearance is caused by waxes that naturally occur on the outside of the corn kernel. These can be easily removed by winterization, i.e., cooling the oil to a low temperature and filtering off the insoluble material.

The final step in oil refining is deodorization. The oil is held at a high temperature (218–235°C) under reduced pressure. The relatively volatile constituents that are responsible for the odors and flavors are removed, as well as the free fatty acid and colored materials. The vacuums used are very high, and oxygen must be rigidly excluded at those temperatures. Peroxides and aldehydes are removed by the process. The oil is then ready to bottle. Typical analytical data for corn oil would show a free fatty acid content (as oleic acid) of 0.20%, a saponification value of 190, an iodine value of 125, and a smoke point of 232°C.

REVIEW QUESTIONS

1. What are the products of a wet corn mill?
2. Why is sulfur dioxide added during steeping?
3. What happens to the solubles from a corn wet mill?
4. What is a hydroclone?
5. What principle is used to separate the starch from the protein in corn wet milling?
6. What is meant by the statement "the system is bottled up"?
7. What are the disadvantages of using sorghum as a starting material rather than corn?
8. What is generally the starting material to make wheat starch?
9. What is the difference between the A and B starches of wheat?
10. Why is sulfur dioxide not used in a wheat wet-milling operation?
11. What is used to free the starch from the protein in rice wet milling?
12. What cereals are used to produce oil?

13. From what part of the kernel is rice oil obtained?
14. Why does corn oil generally sell for a higher price than other oils?
15. What is soap-stock?
16. Explain how free fatty acids and phospholipids are removed from oil.
17. How is oil bleached?
18. What is winterization?
19. How is oil deodorized?

SUGGESTED READING

ANDERSON, R. A. 1970. Corn wet milling industry. Pages 150-170 in: Corn: Culture, Processing and Products. G. E. Inglett, ed. Avi Publishing Co., Westport, CT.

BERKHOUT, F. 1976. The manufacture of maize starch. Pages 109-134 in: Starch Production Technology. J. A. Radley, ed. Applied Science Publishers, London.

FELLERS, D. A. 1973. Fractionation of wheat into major components. Pages 207-228 in: Industrial Uses of Cereals. Y. Pomeranz, ed. Am. Assoc. Cereal Chem., St. Paul, MN.

HAMER, R. J., WEEGELS, P. L., MARSEILLE, J. P., and KELFKENS, M. 1989. A study of the factors affecting the separation of wheat flour into starch and gluten. Pages 467-477 in: Wheat is Unique. Y. Pomeranz, ed. Am. Assoc. Cereal Chem., St. Paul, MN.

JULIANO, B. O. 1984. Rice starch: Production, properties and uses. Pages 507-528 in: Starch Chemistry and Technology, 2nd ed. R. Whistler, J. N. BeMiller, and E. Paschall, eds. Academic Press, Orlando, FL.

KNIGHT, J. W., and OLSON, R. M. 1984. Wheat starch: Production, modification, and uses. Pages 491-506 in: Starch Chemistry and Technology, 2nd ed. R. Whistler, J. N. BeMiller, and E. Paschall, eds. Academic Press, Orlando, FL.

MATZ, S. A. 1970. Starch and oil production from cereals. Pages 300-337 in: Cereal Technology. S. A. Matz, ed. Avi Publishing Co., Westport, CT.

MAY, J. B. 1987. Wet milling: Process and products. Pages 377-397 in: Corn: Chemistry and Technology. S. A. Watson and P. E. Ramstad, eds. Am. Assoc. Cereal Chem., St. Paul, MN.

ORTHOEFER, F. T., and SINRAM, R. D. 1987. Corn oil: Composition, processing, and utilization. Pages 535-551 in: Corn: Chemistry and Technology. S. A Watson and P. E. Ramstad, eds. Am. Assoc. Cereal Chem., St. Paul, MN.

REINERS, R. A., and GOODING, C. M. 1970. Corn oil. Pages 241-261 in: Corn: Culture, Processing, and Products. G. E. Inglett, ed. Avi Publishing Co., Westport, CT.

WATSON, S. A. 1984. Corn and sorghum starches: Production. Pages 417-468 in: Starch Chemistry and Technology, 2nd ed. R. Whistler, J. N. BeMiller, and E. Paschall, eds. Academic Press, Orlando, FL.

WEBER, E. J. 1973. Structure and composition of cereal components as related to their potential industrial utilization. IV. Lipids. Pages 161-206 in: Industrial Uses of Cereals. Y. Pomeranz, ed. Am. Assoc. Cereal Chem., St. Paul, MN.

Rice, Oat, and Barley Processing

CHAPTER 8

The three grains discussed in this chapter are similar in that all are harvested with the hulls attached. The other cereals lose their hulls during the threshing step and are handled as naked grains. As might be expected, the first step in processing these cereals is to remove the hulls. Barley's hull tightly adheres (in fact, is "cemented") to the outer layers of the pericarp. The hull of rice or oats is, instead, a separate intact structure that essentially surrounds the grain with no bonding between the grain and hull.

Rice

Rice with the hull on is called *paddy* or *rough rice*. About 20% of paddy rice is hull. The kernel remaining after the hull is removed is *brown rice*.

The most common huller used today is the *rubber-roll sheller* made in Japan (Fig. 1). The rough rice is passed between two rubber-coated rolls that turn in opposite directions and are run at a differential (different speeds). The pressure and shear remove the hulls much as rubbing peanuts between your hands removes the shells. The pressure exerted by the rolls can be varied, as different cultivars may require different pressures for adequate shelling. Excessive pressure may discolor the grain and cut down the life of the rolls, which is already limited. The rolls must be replaced every 100–150 hr.

Rubber-rolled shellers are preferred because of their efficiency in removing hull (>90%) and because they cause less breakage than the older types of shellers. One older type is the disk sheller, which has a horizontal abrasive-surfaced wheel that rotates on a vertical axis just below a stationary abrasive-surfaced wheel (Fig. 2). Rough rice enters through an opening in the center of the stationary top wheel and, because of the spinning bottom wheel, brown rice flows out between the two wheels. The abrasive action of the wheels on the rice hull essentially sands off part of the hull and frees the brown rice. This type is less efficient and results in more breakage.

After separation, the hull is removed by aspiration, and the remaining rough rice is separated from the brown rice. The separation, which is based on bulk density, can be made on a gravity separator (sometimes called a *paddy machine*, see Fig. 6 of Chapter 6). The paddy is returned for another pass through the sheller. Products at this point are hulls, brown rice, and broken brown rice.

It is common knowledge that brown rice is more nutritious than white (milled) rice, but hardly anyone eats brown rice. The last sentence is undoubtedly true. The reason behind this situation could be debated, but, in general, we eat not for nutrition but for other, complex reasons.

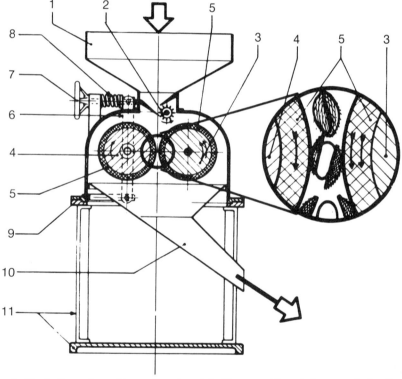

Fig. 1. Illustration of a rubber-roller rice dehulling machine: 1, hopper; 2, feeding roll; 3 and 4, drive rolls; 5, rubber coating; 6–8, pressure adjustment system; 9, housing; 10, delivery spout; 11, stand.

For one thing, the cooking time for brown rice is much longer than that for white rice. Undoubtedly a strong relationship exists between the preferences for eating white or brown rice and those for eating white or brown bread.

The milling of brown rice is essentially the removal of the bran by pearling. Dry calcium carbonate (about 3.3 g/kg) is added to the brown rice. The calcium carbonate is an abrasive that helps in removing the bran. In some cultivars of rice, the bran adheres to the grain more tightly than in others, and a small amount of water may be added to soften the bran layers.

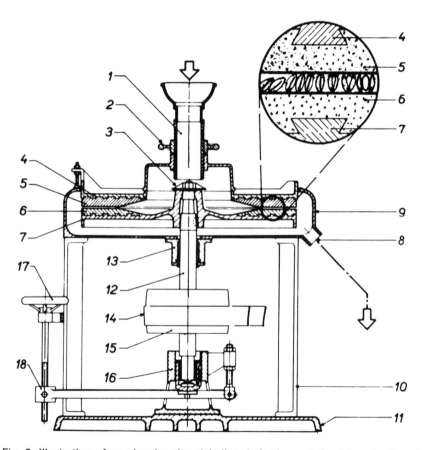

Fig. 2. Illustration of an abrasive rice dehuller: 1, feed spout; 2, slide gate; 3 and 7, rotating abrasive wheels; 4, fixed wheel; 5 and 6, abrasive stones; 8, outflow spout; 9, housing; 10 and 11, stands; 12–16, drive mechanism and bearings; 17 and 18, adjustment system for distance between stones.

The pearler or milling machine is the most critical machine in a rice mill (Fig. 3). This is the machine in which bran is removed and also in which most of the breakage occurs.

In a typical pearler, the brown rice enters through a flow-regulating valve and is then conveyed by a screw to the pearling chamber, where the mixing roller causes the grains to rub against each other, abrading off the bran. Most of the bran is removed by the grain rubbing on other grains, although a small amount is also removed by the grain rubbing on the steel screen surrounding the chamber. At the discharge end of the pearling chamber is a plate held in place by a weight. The position of the weight varies the pressure on the plate and thus the back pressure on the rice in the pearling chamber. The degree of milling can be controlled by varying the pressure and thereby changing the

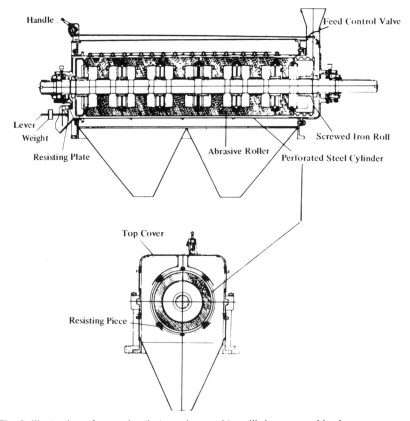

Fig. 3. Illustration of a pearler that can be used to mill rice or pearl barley.

average residence time in the chamber. Generally, the miller visually determines the proper setting and the degree of milling.

The miller tries to achieve a reasonable throughput, with a uniform degree of milling, and at the same time keep breakage to an absolute minimum. The *head rice* (unbroken milled kernels) brings a much larger price than the brokens, which are generally sold to brewers to be used as adjunct. It has been estimated that a 1% change in breakage can cause a $100,000 difference in profit to an average-sized rice mill.

After milling, the loose bran is removed by an aspirator. The milled rice is then polished. The polisher consists of a rotating vertical cylinder to which straps of leather are attached. The milled rice passes downward between the rotating cylinder and the surrounding wire screen. An additional amount of bran is removed by the polisher. Some mills have discontinued the use of polishers because of the increased breakage produced. After being polished, the head rice is separated from brokens by screening or by disk separators.

A rice milling scheme is outlined in Fig. 4. The products from the mill are head rice, brokens, rice bran, rice polish, and hulls. In general, paddy rice yields 20% hulls, about 8% bran, and about 2% polish. The remaining 70% is brokens or head rice. The amount of brokens varies widely and depends on the rice cultivar, the milling scheme, and the skill of the miller.

RICE HULLS

Rice hulls are mostly a nuisance to the miller. They are tough, fibrous, abrasive, and have low nutritive value. They consist of two interlocking halves; the larger is the lemma, the smaller the palea. The interlocking of the two is what makes removal of the hull from the kernel difficult.

Proximate analysis of rice hulls shows them to be high in ash ($\sim$20%), cellulose ($\sim$30%), pentosans ($\sim$20%), and lignin ($\sim$20%) and to contain smaller amounts of protein ($\sim$3%) and fat ($\sim$2%). In addition, they contain small amounts of vitamins.

The predominant component (94–96%) of the ash from rice hulls is silica. The uptake of silica, in all cereals except rice, is passive. That is, the soil solution contains a certain level of soluble silica, and that silica is taken into the plant along with the water. Because plants do not have an elimination system, whatever is taken in must be deposited in the plant. This fact has been used to determine the water requirements of plants. Rice, however, is different. It actively takes up silica (that is, it takes up more silica than would be calculated from the water uptake). Why the rice plant takes up these amounts

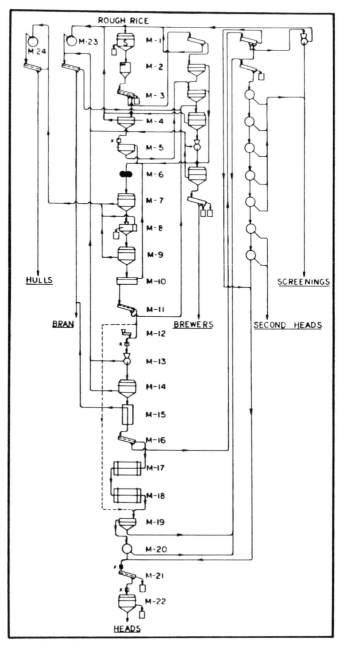

Fig. 4. Flow of a rice mill.

of silica is unknown; however, silica has been related to disease resistance in rice. Whatever the reason, the silica must be deposited somewhere in the plant, and the hull is the depository for much of it.

Numerous uses for rice hulls have been suggested in the literature. However, the disposal or utilization of them still remains a major problem. Estimates indicate that, of the hulls produced, as much as one third is not utilized. Early workers felt that hulls were dangerous in feed, but this has been disproved. The most common use of hulls is to mix them with rice bran and sell them as rice millfeed. Such products generally contain about 61% hull, 35% bran, and 4% polish.

Hulls can also be used as bedding or litter, as a fertilizer or mulch, and in a number of industrial uses, including burning for fuel. They are essentially sulfur free. The heat produced may be two-thirds that of wood. Small amounts of hulls are used to produce carbon and also furfural. The rice hull yield of furfural, an important organic intermediate, is low compared to the yield from such sources as oat hulls, cottonseed hulls, or corn cobs and is generally related to the pentosan level. Small amounts of hulls are also used as filter aids or abrasives.

RICE BRAN AND POLISH

Rice bran and polish are by-products from rice milling. The bran is the outer layers of the pericap from brown rice; the polish is the inner layers, containing aleurone cells and small amounts of starchy endosperm. As might be expected, the amounts of bran and polish vary widely, depending on the milling procedure employed. As stated earlier, bran normally is about 8% of brown rice and polish about 2%. The general composition of the two fractions is shown in Table I. In addition

TABLE I
Composition of Rice Bran and Polish[a]

Constitutent	Bran	Polish
Protein	12.0	12.0
Fat	13.0	16.0
Ash	10.0	8.0
NFE[b]	40.0	56.0
Crude fiber	12.0	7.3
Pentosans	10.0	...

[a] Adapted from D. F. Houston. 1972. Pages 272-300 in: Rice: Chemistry and Technology, 1st ed. D. F. Houston, ed. Am. Assoc. Cereal Chem., St. Paul, MN.
[b] Nitrogen-free extractives.

to the values shown, rice bran ash is high in magnesium, potassium, and phosphorus. Rice bran has also been shown to reduce cholesterol.

Bran is also an excellent source of B-vitamins (typical values are thiamin, 10.6; riboflavin, 5.7; niacin, 309; and pyridoxine, 19.2 $\mu g/g$) and vitamin E but contains little or no vitamins A, C, or D.

The utilization of rice bran and polish is not as great as one might expect from their composition. Both bran and polish are excellent sources of nutrients in animal feeds. However, when rice bran is stored without inactivation of lipase, the fat in the bran rapidly becomes hydrolyzed and oxidized, causing the bran to become rancid and unpalatable. The bran, particularly if it is damp, is an excellent growth media for microflora; the microflora can produce mycotoxins and bacterial toxins. When either of those conditions prevails, the bran cannot be used as a feed and therefore is used as fuel or fertilizer.

Rice bran is a source of oil. Generally, the bran is solvent-extracted. It contains appreciable amounts of wax that must be separated from the oil. After refining, rice oil is comparable to other edible oils.

RICE QUALITY

As with most food products, the quality of rice is determined by its ability to produce the desired end product. Because most rice is consumed as whole kernels, the milling yield of head rice is an important quality factor. Consumer acceptance varies from country to country and even between regions within a specific country. Most U.S. consumers prefer rice that cooks to produce dry and fluffy kernels. Each kernel should retain its shape and separate identity after cooking. Other consumers, particularly in Asia, prefer rice that cooks to be moist and chewy, with the kernels sticking together.

Rice cultivars grown in the United States are divided by grain size and shape into three types: short-, medium-, and long-grained (Fig. 5). Typical long-grain types are dry and fluffy after cooking and are preferred for quick-cooking rice, canned rice, canned soups, and convenience foods containing rice. Typical short- and medium-grain types cook moist and sticky and are suitable for breakfast cereals, baby foods, and brewing.

In the United States, all rice is bland in taste and translucent in appearance. A small amount of waxy rice (with all-amylopectin starch) is grown; it generally has an opaque endosperm. In other parts of the world, many rice cultivars are scented (aromatic). These usually yield poorly, but in many countries small areas are grown by farmers for their personal use.

Rice varies rather widely in a number of physicochemical properties. This is in contrast to other grains. For example, the amylose content of rice starch varies widely, depending upon the cultivar and type. U.S. long-grain cultivars are characterized by higher amylose content (23–27%) than are short- and medium-grain types (15–21%). These differences in amylose content greatly change many other properties of the rice. However, recent data suggest that the amylose content may not actually vary. The apparent amylose content does vary, but this may actually be because of differences in the amylopectin structure.

Rice endosperm varies in its composition more than do the endosperms of other cereals. Flour abraded from the outer layer of milled rice may contain 20% protein, whereas the total milled kernel may contain only 8%. Similar results have been reported for lipids (6% compared to 0.3%), vitamins (four to eight times higher in the outer layers), and certain minerals. The starch content, of course, changes in the opposite direction from the protein content: that is, starch content is higher in the center of the endosperm and lower at the outer edges. This type of data must be examined carefully, as a small amount of aleurone left on the outside of the milled kernel could alter the data significantly.

Fig. 5. Rice: short- (top), medium- (middle), and long-grain (bottom).

AGING OF RICE

Like all cereals, rice undergoes *after-ripening*, i.e., biochemical changes that occur in the dry sound grain as a function of storage time after harvest. The phenomena is not well understood, but the changes it causes are easily shown. Rice milled from freshly harvested paddy gives very pasty and sticky kernels after cooking. If the paddy is stored under good storage conditions, within weeks, the milled grain will cook with much less tendency to stick together. Storage of milled rice also produces decreased cohesiveness, drier surfaces, larger volumes, and firmer texture in cooked kernels. In addition, cooking time tends to become longer with increased storage time.

For consumers preferring cohesive rice (e.g., in Japan, Korea, Taiwan, and China), those changes brought about by after-ripening are undesirable. However, if you prefer your rice dry and fluffy (as do most people in the United States, Italy, France, Spain, and India), then these are desirable changes.

Although many changes can be demonstrated in physicochemical properties and enzyme activities (generally a decrease in activity) with aging, we do not yet know which of these changes results in the changes in cooking quality described above.

QUICK-COOKING RICE

Milled rice requires 20–35 min to cook. This relatively long time is caused by the slow rate at which water diffuses into the kernel. Most rice is translucent and thus tightly packed, with no air space or other channel for water to penetrate the kernel. For cooking to occur, water must penetrate to the center of the kernel with sufficient heat capacity to gelatinize the starch. Thus, to produce a quick-cooking rice, we need to provide channels for water to penetrate the kernel during cooking.

Several techniques are used to produce quick-cooking rice. One is precooking to about 60% moisture (fully cooked rice contains about 80% moisture), followed by slow drying to 8% moisture; this process allows the rice to maintain a porous structure. Another is rapid heating of dry rice at about 10% moisture, which causes internal fissuring. Pregelatinization, as described above, and rolling or bumping the grains followed by drying produces a relatively flatter grain (which means that water has less distance to diffuse during cooking). Precooking followed by freezing, thawing, and drying or puffing by rapid changes in pressure have also been used to produce quick-cooking rice. In general, anything that opens the kernel and allows the water to penetrate readily decreases cooking time. As a result of most of the methods used to produce quick-cooking rice, the uncooked kernels become opaque and larger in volume.

PARBOILED RICE

Parboiling, the process of heating paddy rice in water and then drying it, has been practiced since ancient times. It was probably started to aid in dehusking (dehulling). Today we know that its major advantage is nutritional. As mentioned earlier, much of the vitamin and mineral content of rice is concentrated in the outer layers. Parboiling helps to move these nutrients into the kernel.

It is estimated that as much as 25% of world rice is parboiled. Its major consumption is in India, Sri Lanka, and parts of Africa. Very little parboiled rice is consumed in Asia.

Parboiling consists of three steps; steeping, steaming, and drying. Steeping is generally done at about 60°C. Lower temperatures require longer times, and thus the chances of fermentation, sprouting, and other side effects are greater. After the steeping, the excess water is removed, and the paddy is heated with steam to gelatinize the starch and sterilize the paddy. The gelatinization of the starch traps the vitamins in the endosperm. The rice can be rapidly dried to about 18–20% moisture and then dried more slowly (this avoids cracking and fissuring).

Parboiled rice is not quick-cooking. In fact, it requires slightly longer cooking times. Parboiled rice has advantages and disadvantages relative to its regularly milled counterpart. The major advantages are: a higher yield of head rice from milling because the kernel is more resistant to breakage, more resistance to insects, more nutritiousness (particularly more vitamin B_1), and less tendency to become sticky or mushy during cooking. This, of course, explains why the process is not popular in Asia, where consumers do want a sticky rice. The major disadvantages are a darker color, a slightly different flavor, and increased susceptibility to rancidity.

RICE ENRICHMENT

The consumption of a large part of one's diet as white milled rice can lead to a deficiency in the B-vitamin thiamin, leading to the disease beriberi. Parboiling reduces the incidence of the disease but has its own disadvantages. In the United States, parboiling is not a replacement for enrichment.

Rice is more difficult to enrich than are cereal flours. The most popular method of enriching rice in the United States is to add powdered nutrients. However, these nutrients are often washed off by the consumer, and in many parts of the world, rice is cooked in excess water and the cooking water is discarded. Another method of enriching rice is to produce a rinse-resistant premix and coat the rice with it. Milled rice is tumbled in a slowly rotating polishing cylinder and sprinkled with an acidified aqueous solution of thiamin and niacin. The moisture is removed by aspiration, and a coating of stearic acid, zein, and abietic acid in alcohol is sprayed on the rice. The mixture of the latter three gives a rinse-resistant coating. The rice is then dusted with ferric pyrophosphate (to supply iron).

In Japan, other approaches to enrichment have been used, for example, the addition of water-insoluble derivatives of thiamin (benzoyl thiamin disulfide). Another approach is to produce simulated rice kernels made from cereal flour doughs that contain the enrichment. Disadvantages of the latter technique include the appearance of the kernels (if they look different, the consumer will pick them out and discard them) and the fact that if they are made too inert, they may pass the intestine unutilized.

Oat Milling

Most of the oats grown in the United States are used for feed for poultry or other animals. Only about 10% of the crop is processed for human consumption. Oats are classified by color; milling oats are white. In the United States, most oats and all milling oats are grown in the north central part of the country. This is reasonable as oats prefer a cool, moist climate.

Oats are harvested with their hulls on. The hulls represent about 25% of the total weight. After they are removed, the oats are called groats. Visually, these resemble wheat or rye kernels except that the germ contains a long scutellum and the kernel is covered with *trichomes* (hairs). Oat groats are higher in fat and protein than are most other cereals. They are also a good source of several enzymes. The most troublesome of these is very active lipase. Unless the lipase is denatured, milled products have a very short shelf life because of the production and subsequent oxidation of fatty acids.

When milling oats are received at the mill, the first step is a thorough cleaning. Besides removing the usual foreign material (seeds, sticks, and dirt), the cleaning also must remove oats that are unsuitable for milling. These include *double oats*, those in which the hull (primary envelope) also covers a secondary grain; the groats of both the primary and secondary grains are usually poorly developed, resulting in a high percentage of hull. Also removed are *pin oats* (usually very thin with little or no groat) and *light oats* (of normal size but with little or no groat) (Fig. 6). All these types are removed and sold as feed material.

After cleaning, the oats are heat-treated or dried. The heat treatment generally consists of heating the oats for about 1 hr in large open pans heated with steam. The oats reach a temperature of about 93°C and lose 3–4% moisture. As a result of the treatment, they take on a slightly roasted flavor that is considered desirable. In addition to the flavor changes, drying makes the hulls more friable and thus easier to remove. It also inactivates the lipolytic enzymes in the kernel. The denaturation of those enzymes is critical if one is to produce products with good shelf life.

The heat-treated oats are then graded for size and dehulled. This is done by disk separators dividing them into large and stub, or short, oats. Both grades are dehulled, but not together. The most common dehuller in use now is the impact huller (Fig. 7). Oats enter the center of a high-speed rotor, which throws the oats against a rubber liner fixed to the outside case of the machine. The rubber liner reduces

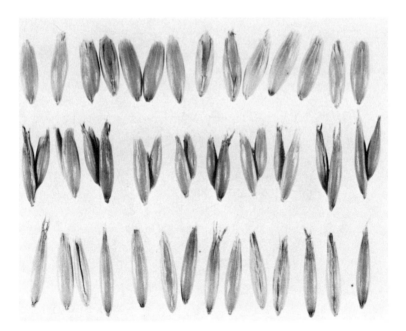

Fig. 6. Oats: regular (top), double (middle), and pin (bottom).

breakage and assists in separation of the hull from the groat. The hulls that are freed by the huller are light enough to be removed by aspiration. Care must be taken not to remove small chips of groats with the hulls, as this lowers the yield. The groats are then removed from the unhulled oats by sieves or disk separators. Paddy separators (see the section on rice) can also be used to separate oats destined for another pass through the dehulling machine. Paddy machines are quite effective in making the separation but have very limited capacity.

For the production of old-fashioned *rolled oats*, groats free of hulls must be used. To produce the rolled oats, the groats are steamed and rolled immediately afterwards. Steaming (with live steam at atmospheric pressure) just before flaking accomplishes several things. It makes the groats more flexible so that fewer of them are broken during the flaking operation. It also helps to denature enzymes that cause rancidity if not inactivated.

To produce quick-cooking oats (the 1-min kind), the groats are cut and then steamed and flaked (rolled). The purpose of cutting is to produce three to four uniform pieces from each groat. The cutters consist

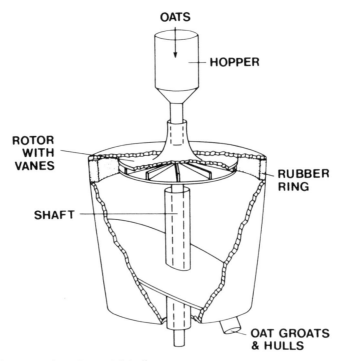

Fig. 7. Cut-away view of an oat dehuller.

of rotating perforated drums. The groats position themselves endwise in the perforations and are cut by stationary knives at the outside surface of the drum. The cut groats are steamed and flaked in a process similar to that used with the whole groats. The rolled flakes produced from the cut groats are much thinner, and water must diffuse a much shorter distance during cooking. Thus, the thin flakes are quick-cooking. To produce instant-cooking flakes, they must be thinner yet. The thinner the flake, the faster it cooks, but it also tends to lose its identity much faster. This makes the quick or instant product become mushy much more quickly.

After flaking, the flakes are cooled with air. The air stream helps to remove any hull slivers that have escaped the previous efforts of removal and have become loose during the cutting and flaking operations. The flakes are then packaged in breather-type packages (packages that allow air movement through them). The breather-type package is necessary to allow air to remove rancid odor produced during storage. Even though the enzymes are denatured, a small amount of rancid odor is produced. However, a reasonable shelf life can be obtained if the product is allowed to breathe.

The products of oat milling are the rolled flakes (of various thicknesses); hulls (about 25% of the oats), which are used as a high-fiber feed ingredient or for furfural production; and oat flour, produced as a by-product of the cutting and flaking operations. This last is finding increased use in baby foods and breakfast cereals.

Barley Pearling

The amount of barley that is pearled is very small compared to the amounts used for feed or for malting. The hull of barley is strongly attached to the pericap. Thus, barley is difficult to dehull by the techniques used for rice or oats and generally is *pearled*. Pearling is abrading away the outer surfaces of the grain with an abrasive surface—sanding the grain, if you will, to remove the hull and the pericap. Pearled barley is a common ingredient in many soups. The pearled grain can also be reduced to a flour, which is used in baby foods or breakfast cereals.

REVIEW QUESTIONS

1. How is the hull of barley different from the hulls of rice and oats?
2. What is paddy and what is another name for it?

3. How is rice dehulled?
4. How is paddy separated from brown rice?
5. Why is calcium carbonate added during rice milling?
6. How does a rice milling machine work?
7. What is head rice?
8. What is polishing and how is it accomplished?
9. How is head rice separated from brokens?
10. The ash of rice hulls is high in what mineral?
11. How are rice hulls utilized?
12. What is the major problem in utilization of rice bran?
13. What is the most important factor in rice milling?
14. What are the important factors controlling rice quality?
15. Explain after-ripening in rice.
16. How is quick-cooking rice produced?
17. What is parboiled rice and how is it accomplished?
18. What are the advantages and disadvantages of parboiled rice?
19. How is rice enriched?
20. What percentage of oats is hull?
21. Why are oats heat-treated before they are milled?
22. How are groats removed from oats?
23. How are quick-cooking oats produced?
24. How is barley milled?

SUGGESTED READING

BECHTEL, D. B., and POMERANZ, Y. 1980. The rice kernel. Pages 73-113 in: Advances in Cereal Science and Technology, Vol. 3. Y. Pomeranz, ed. Am. Assoc. Cereal Chem., St. Paul, MN.

BHATTACHARYA, K. R., and ALI, S. Z. 1985. Changes in rice during parboiling and properties of parboiled rice. Pages 105-168 in: Advances in Cereal Science and Technology, Vol. 7. Y. Pomeranz, ed. Am. Assoc. Cereal Chem., St. Paul, MN.

KELLY, V. J. 1985. Rice in infant foods. Pages 525-537 in: Rice: Chemistry and Technology, 2nd ed. B. O. Juliano, ed. Am. Assoc. Cereal Chem., St. Paul, MN.

SALISBURY, D. K., and WICHSER, W. R. 1971. Oat Milling. Systems and Products. Assoc. Oper. Millers Tech. Bull. (May) pp. 3242-3247.

VAN RUITEN, T. L. 1985. Rice milling: An overview. Pasges 349-388 in: Rice: Chemistry and Technology, 2nd ed. B. O. Juliano, ed. Am. Assoc. Cereal Chem., St. Paul, MN.

YOUNGS, V. L., PETERSON, D. M., and BROWN, C. M. 1982. Oats. Pages 49-105 in: Advances in Cereal Science and Technology, Vol. 5. Y. Pomeranz, ed. Am. Assoc. Cereal Chem., St. Paul, MN.

Malting and Brewing

CHAPTER 9

The cereal most often malted is barley, although sizable quantities of wheat and rye are also used, and in parts of Africa, sorghum is malted. Occasionally, references are found in the literature to malting of millet, particularly finger millet. In theory at least, any cereal could be used. However, the type and amount of enzymes produced vary from one cereal to another.

The predominant use of barley is undoubtedly based on a number of factors. To start with, it has traditionally been the cereal of choice. In addition, it gives a good balance of enzymes and has tightly adhering hulls that protect the modified grain after malting and provide a natural filter bed later in the brewing process.

Simply stated, *malting* is the controlled germination followed by the controlled drying of a seed. The goal is to produce high enzyme activity, endosperm modification, and a characteristic flavor with a minimum loss of dry weight. Grain quality is important to producing good malt. Grain selected for malting must be sound and plump, have a high capacity for germination, be free of trash and broken kernels, and be relatively free of molds.

Dormancy and After-Ripening

Dormancy and after-ripening are two major problems associated with malting. True *dormancy*, defined as sound seeds that will not germinate under suitable conditions, precludes malting. Dormancy is found in all seeds to some degree. If seeds were not dormant at the time of ripening, they would sprout while still in the head.

The study of the mechanisms and duration of dormancy in various seeds is fascinating. For example, the seeds of certain trees are impregnable by water and must go through several years of freezing and thawing (or a forest fire) before they will take up water and then germinate. Dormancy of certain tree seeds can sometimes be broken by boiling the seeds in sulfuric acid! There are other types of dormancy. Some seeds (e.g., lettuce) need light to break dormancy and therefore must be planted on top of the soil. Another type is found in seeds of desert plants that contain a water-soluble germination inhibitor. These will not germinate until sufficient water is available both to dilute the inibitor to where it is not effective and to allow the plant to complete its life cycle.

Cereal grains, in general, have relatively low levels of dormancy. Generally, within a few weeks of harvest, their dormancy has disappeared (been "broken"). In fact, many times the major problem with cereals is a lack of dormancy, which allows the grain to sprout prematurely in the field. This gives high enzyme activities and lower-grade products. The problem is particularly bad in northern Europe, where the weather is often wet and humid during harvest. In the United States, it is a rare year when sprouting is not a problem in some areas. In wheat, a genetic link appears to exist between resistance to sprouting (dormancy) and seed color; red wheat tends to have greater resistance to sprouting than does white wheat.

After-ripening is a collective term used to describe the biochemical changes that occur in a seed after it is mature, naturally dried, and harvested. One must remember that a healthy seed, even though it contains only 12% moisture, is alive and continues to undergo biochemical changes after harvest.

We do not understand the biochemical changes that occur after harvest but are well acquainted with many of their consequences. Barley that is malted soon after harvest produces a poor wort (characterized by a lower soluble extract and a hazy appearance). The malt from barley that has been aged for three months after harvest does not give these problems. In practice, new-crop barley is not malted until it has been stored for about three months.

After-ripening is not restricted to barley but appears to be a general phenomenon of all cereals. For example, as rice ages, it swells more during cooking, loses less solubles to the wash water, and is not as sticky as "new" (freshly harvested) rice. In India, a significant price differential is often found between old and new rice. The phenomenon is also seen in wheat, with "new" wheat giving flour that does not perform satisfactorily. It is well known that every year, as processing of new-crop grain is started, the problems are large. After two to three months of large problems, everything settles down and the crop can be processed. A large amount of the yearly bout with the new crop could probably be avoided if the crop were stored for two to three months and then processed. Of course, this would be expensive.

The Malting Process

The malting process is started with a rather rigorous cleaning of the barley to remove all foreign seeds, broken kernels, etc. In addition to cleaning, the barley is graded into three classes: thin, plump, and very plump. The thin is sold as feed, and both the plump and very plump are malted, but not together.

STEEPING

Steeping, the process of soaking the grain in water, has the primary purpose of introducing water into the grain. The grain's moisture content is increased from about 12 to 44%. Another factor occurring during steeping is the leaching of components from the hull. These are discarded in the steep water. During steeping, germination is started. Steeping is generally complete when the moisture content of the grain is raised to 42–44%. The 42–44% moisture level is an equilibrium value. It is the point at which the hydrostatic pressure in the cell equals the osmotic pressure generated by the cell sap.

It is important that the moisture penetrate to the center of the kernel. Because diffusion is the mechanism by which this occurs, steeping is a slow process. The time required depends on the distance the water must diffuse, and therefore more time is required for a very plump kernel as opposed to just a plump kernel. This, of course, is why the grain is graded for plumpness before steeping. The other major factor affecting steeping time is the steeping temperature. Higher temperatures give shorter steeping times because water diffusion is faster. However, the higher temperature also promotes the growth of microorganisms. Steeping is generally done at about 15°C.

If the steeping time is too short, the barley is *understeeped*. This means that insufficient water has been absorbed. The center of the kernel is often dry, and generally the resultant malt is of low quality.

Oversteeping, on the other hand, is not the uptake of too much water; after equilibrium is reached, no additional water is absorbed. Oversteeping appears to be related to the extended time of submersion of the grain. The results of oversteeping are delay in the onset of germination, increased growth of mold and bacteria, and the production of undesirable odors.

A number of steps are taken to protect against oversteeping. The steeping grain is aerated either continuously or intermittently. If air is not provided, barley respiration uses the available oxygen and creates an aerobic condition. Providing air stops anaerobic respiration of molds and removes CO_2 from the water. CO_2 lowers the pH, which encourages the growth of bacteria. The pH of the water may be adjusted with lime. The grain may also be agitated or turned. This discourages the buildup of bacteria between kernels that are touching. The pH of the water may be adjusted with lime. The steep water may be frequently or continuously changed. In general, all of the above are designed to keep microorganisms under control and stop the production of undesirable odors.

GERMINATION

After steeping is complete, the grain is removed from the water and placed in beds to germinate. Physiologically, *germination* is the process by which a new plant starts to form (Fig. 1). Both rootlets and an acrospire are formed. Germination usually takes four to five days. During that time, moist air is forced continuously through the germinating bed of grain to keep the relative humidity close to 100%. The temperature is generally held at about 15°C, and the grain bed

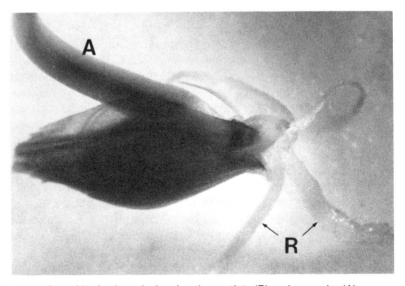

Fig. 1. A germinated barley kernel, showing the rootlets (R) and acrospire (A).

is usually "wet down" two to three times during the germination period. Water at the rate of about 1.5 gal/bu is used.

The growth of the new plant is controlled by the moisture of the grain, the temperature of the grain, and the amount of air forced through the grain bed. The goal is the minimum amount of growth that will obtain maximum endosperm modification and yield of malt of high enzyme activity. Germinating at higher temperatures would give a higher rate of growth of the new plant, but the enzyme activity and yield of malt would both be lower. As a rule of thumb, the germinating process for malting is considered to be complete when the acrospire is one-third to three-fourths the length of the kernel.

KILNING

After the grain has germinated, the green malt (not green in color but green in the sense of green wood, i.e., undried) has about 45% moisture. It is then dried to halt growth, give a storable product, and develop the characteristic flavor and color of malt. The drying and flavor development process is called *kilning*. During germination, as the new plant was starting to develop, a large number of enzymes were synthesized or activated. Generally, when we think of the enzymes of malt, we think of α- and β-amylases. These are undoubtedly very important enzymes in the malt system, but they are only two of a large number of enzymes present. The problem is to dry the green malt in such a way as to remove the moisture without damaging the enzyme activity.

At high moisture contents, many enzymes are sensitive to heat. The general relationship of protein solubility to temperature for various moisture contents is shown in Fig. 2. Enzyme activity is, in general, affected the same way as protein solubility is. Thus, to protect enzymes, we must heat the green malt carefully and with low temperatures. After the malt is reduced to a relatively low moisture, the temperature can be increased to bring about the browning reaction that gives the malt its characteristic flavor and color. By varying the amount and severity of drying and browning, a range of malt colors and flavors can be produced. Specialty malts with dark color are produced and used to produce dark beers. Other malts are dried with low temperatures to produce "green" (not browned) malts that are stable and have high enzymatic activities. In general, there is a trade off between enzyme activity and flavor and color. If you want high enzyme activity, it is produced at the expense of flavor; conversely, to create high flavor, you must dry so severely that a great deal of enzyme activity is lost.

During malting, a 10–20% loss of weight occurs. The loss is caused by the growth of rootlets that are removed during the malt cleaning process. In addition, a part of the endosperm is consumed to provide the energy to drive the germination process.

Some malt is bleached by being *sulfured* (treated with SO_2 at the initial drying stages). In addition to its effect on color, the SO_2 increases the soluble protein in the malt and the activity of proteolytic enzymes. In addition, the SO_2 blocks nitrosamine formation and reduces the bacteria level in the malt.

Malt has a large number of uses. The most important, of course, is in brewing and the production of distilled products. However, other important uses are in the baking industry as an enzyme source and in the breakfast food industry mainly as a flavoring agent.

Brewing

The fermentation of cereal grains to produce beer is as old as recorded history. The principal ingredients used in the production of beer include malt, water, yeast, and hops. Malt serves as both a source of enzymes and a source of fermentable carbohydrate. Traditionally, malt was the only source of carbohydrates, and in Germany it still is today. In the

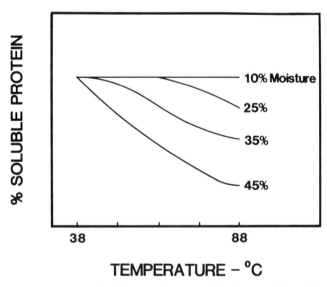

Fig. 2. Change in protein solubility as a function of heating at different temperatures and moisture contents.

MALTING AND BREWING / 183

United States and most of the world, other sources of carbohydrate are used in addition to the malt; these are known as *adjuncts*.

MALT

As mentioned above, malt is the source of enzymes and fermentable carbohydrates. In addition, it makes a major contribution to the flavor and color of the final beer. To produce a dark beer, one must start with a dark malt. For yeast to produce CO_2 and ethanol from cereals, the starch must first be converted to simple sugars (glucose, maltose, and mlatotriose) by the malt enzymes. The enzymes that we are interested in are the amylases, the *diastatic system*. Actually, we are concerned with the ability of a mixture of α- and β-amylases to convert starch to maltose. Sugars larger than maltotriose are not utilized by yeast and thus end up in the beer. Even the maltotriose is utilized very slowly.

Most of the beer in the world is produced with barley malt. The lone exception appears to be the sorghum beer produced in Africa.

ADJUNCTS

Adjuncts are nonmalt sources of potentially fermentable carbohydrates. The trend toward paler and more mildly flavored beers in the United States and other parts of the world has led to the increased use of adjuncts. In addition, strong economic force also favors the use of adjuncts: costs associated with the malting process and the nearly 20% loss of dry material are steep.

The list of potential adjuncts is quite long; virtually any starch source could be included as long as it does not adversely affect the flavor of the final beer. Those that are currently being used include: rice grits, corn grits, unmalted barley, wheat starch, corn syrup, and sorghum grits. Unmalted cereals usually lead to a paler and more mildly flavored beer. The rice consists of kernels broken during milling and then reduced to grits before being used in the brewing process. Corn and sorghum are also milled to grits before use as adjuncts.

A very important specification for grits, other than particle size, is a low level of fat, generally set at less than 1%. The grits must not be rancid, as the flavor will carry over to the beer and adversely affect its quality. Grits for brewing must be made from sound grain that is free of mold and foreign seeds. Because beer is a lightly flavored product, any off-flavor from the adjuncts can be a serious problem.

No matter what is used as an adjunct, the enzymes for conversion of the starch to sugar come from malt. A major part of the carbohydrate also comes from the malt. The major exception to this situation is when high-dextrose-equivalent corn syrup is used as an adjunct. In this case, the carbohydrate is already fermentable by yeast and the amylase is no longer needed.

HOPS

Brewer's hops are the dried fruit (cones) of the perennial plant *Humulus lupulus*. Hops have separate male and female plants, and only the female plant produces the cones. In the United States, hops are grown in the Pacific Northwest and are harvested in late August or early September. The cones must be dried at relatively low temperatures (<50°C) to about 12% moisture. If they are not dried properly, most of the essential oil is oxidized and polymerized.

The brewer is interested in three components of the hops, together called the "brewing principles": the essential oils, the bitter resins, and the tannins. The hop oil is responsible for both aroma and flavor. Its composition is complex, with 70-80% consisting of terpene-like hydrocarbons. However, most of the flavor is thought to come from the oxygenated compounds, a mixture of aldehydes, ketones, alcohols, and carboxylic acids.

The characteristic bitterness of the hops comes from the resinous materials. Two compounds, *humulone* and *lupulone* (Figs. 3 and 4), are peculiar to the hop plant and contribute the bitter taste. During storage, the resins become oxidized and polymerized and the hops lose their bitterness.

The other important components of the hops are the *condensed tannins*, which are polymers of several types of flavonoid compounds. Higher polymers give reddish and brown pigments. Polymers of low

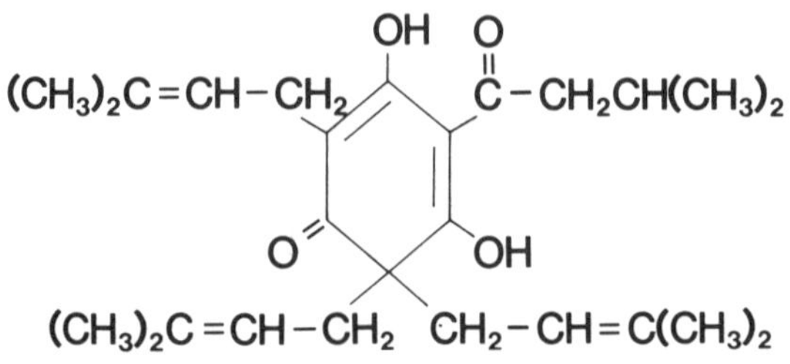

Fig. 3. Structure of lupulone.

molecular weight (mol wt 500–2,000) bind with protein and help make it insoluble. This property, which can be used to "tan" raw hides to leather, gives the compounds their name. It is also the property of the condensed tannins that is important in brewing.

WATER

Even if you only occasionally watch television, you must have learned by now that the best beer can be made only from "sky-blue water" or "Rocky Mountain spring water." There is some truth to those advertisements, as the properties and quality of the water used is of the utmost importance in brewing. The finished beer is about 92% water.

The salts in the water have a pronounced effect on the flavor and character of the beer. In general, a light-flavored beer requires soft water (low in salts), whereas a dark, heavy beer requires hard water (high in salts). Water that is too soft will produce a very flat-tasting beer, as many graduate students have learned from clandestine brewings using distilled water. Water can be treated to remove salts if too hard or to add salts if too soft. For an example of the latter, gypsum ($CaSO_4$) can be added to harden brewing water.

The pH of the water is another important variable but one that can be easily adjusted. Gypsum not only hardens water, it also lowers the pH. Another important factor is the iron (particularly ferric ion) content of the water—levels of less than 1 ppm can cause off-flavors and discolorations. Any odor or flavor in the water is usually carried over into the beer.

YEAST

The yeast used for brewing is usually a strain of *Saccharomyces carlsbergensis*. Some strains of *S. cerevisiae* are also used, particularly

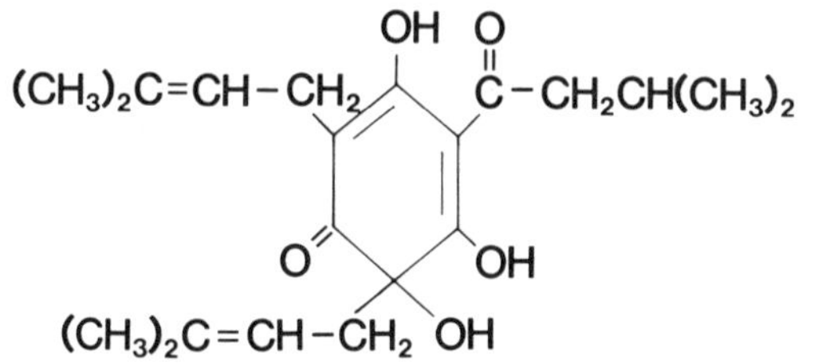

Fig. 4. Structure of humulone.

for ales. A characteristic important to the brewmaster is the vigor with which the strain attacks the fermentable carbohydrate and produces ethanol, carbon dioxide, and the minor constituents that affect flavor.

A second important property of yeast is its *flocculence*, the tendency for the yeast to aggregate or stick together. Individual yeast cells are small enough that they stay suspended in solution for long periods of time, so the rate at which they form aggregates determines how fast they settle from solution. If the cells aggregate and settle to the bottom, the yeast is called a *bottom yeast*. Some yeasts clump together and trap gas, becoming buoyant enough that they are carried to the top; these are called *top yeasts*. A yeast that has a high (rapid) flocculating tendency will be segregated from the fermentable carbohydrates and produce beer with poor *attenuation*. Attenuation is a term describing the degree of conversion of sugar to alcohol. High attenuation results in dry (nonsweet) beers, and low attenuation results in sweet beers. Thus, powder-yeasts (those that do not flocculate well) produce dry beers. However, the powder-yeast is difficult to remove from the beer and, if retained in the beer, gives a yeasty taste. This is a common problem with home-brewed beer, particularly if baker's yeast is used. Thus, the level of attenuation can be used by the brewmaster to regulate the type of beer (dry or sweet) that he or she desires.

The Brewing Process

In the brewing process (Fig. 5), the first step is to reduce the malt's particle size, because when it is received at the brewery, the dried malt still consists of intact kernels, with only the rootlets removed. Either wet or dry milling can be used. In general, the goal is to reduce the endosperm to a fine particle size and still maintain the husk and bran in relatively large pieces. The most popular reduction system is a short-flow roller mill. In this milling system, nothing is removed; all the streams are used in brewing.

The second step is referred to as "adjunct cooking." The adjunct is placed in a cooker with water along with a small part of the malt (Fig. 6). The pH of the mixture is adjusted to about 5.5 (the optimum pH for enzyme activity), and the temperature is held at 35°C for about 30 min. This step, called the *acid rest*, allows the particles of malt to hydrate. The temperature is then raised to about 70°C and maintained for 20–30 min. At 70°C, the starch in the adjunct is gelatinized and therefore becomes much more available to the enzymes in the malt. The small amount of malt is added in the adjunct cooker to thin the gelatinized starch paste. If the starch were not thinned at this point, a much lower ratio of solids to water would have to be used to allow a flowable/pumpable system. After the thinning step, the temperature is raised to boiling for about 30 min. This accomplishes two additional important goals. First, all the adjunct protein is denatured and made relatively insoluble. Protein from the adjunct is not helpful during brewing. The second accomplishment is the sterilization of the system. Thus, in the adjunct cooker, the starch is gelatinized, the resultant mixture is thinned, the adjunct protein is made insoluble, and the mixture is sterilized.

After cooling, the contents of the adjunct cooker are placed in the masher that contains the remainder of the malt slurried in water (Fig. 7). The object of the mashing step is to convert the nonfermentable starch into low-molecular-weight fermentable sugars. The pH of the mash is maintained at about 5.5 during mashing. That pH allows an

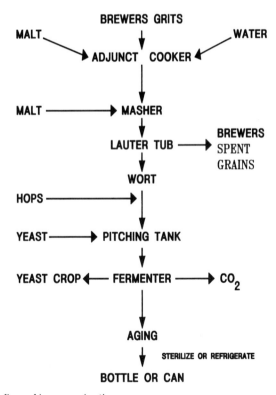

Fig. 5. Simplified flow of beer production.

optimum conversion of starch to sugar by the amylase and also helps to eliminate turbidity caused by incomplete breakdown of protein. Generally, mashing is also a series of programmed temperature rises and holds. Just as with the adjunct cooking step, mashing starts with a 30-min rest at about 35°C, which allows the malt to become hydrated

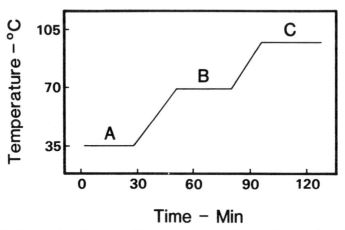

Fig. 6. Outline of the adjunct cooking process. A, acid rest; B, α-amylase thinning; C, sterilization.

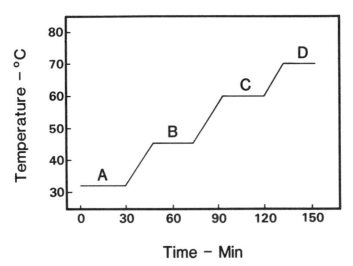

Fig. 7. Outline of the mashing process. A, acid rest; B, proteolytic rest; C, β-amylase optimum; D, α-amylase optimum.

(the malt may be "rested" before it is mixed with the cooked adjunct). The temperature is then raised to about 45°C and held for 20–30 min. This allows enzymes that are susceptible to heat to act. We are primarily interested in the proteolytic enzymes at this temperature. The temperature is then raised to about 60°C and held for another 20–30 min. This is the optimum temperature for β-amylase activity. Mashing is completed after another temperature rise and 20-min hold at 70°C. In addition to the maximum conversion of starch to maltose, modification of the barley protein also takes place during mashing. About two thirds of the malt protein remains insoluble, but the other one third is made permanently soluble and plays an important role at subsequent stages of brewing. If these proteins are not modified, they will cause turbidity and haziness in the beer.

After mashing is complete, the *wort* (soluble material) is separated from the *spent grain*. This is usually accomplished in a lauter tub, which is essentially a large tank with a screen bottom. The mash is pumped into the tub and allowed to settle for about 30 min. The husk and bran particles settle onto the screen and form a filter mat. This is why the husk is left on the barley during malting and also one of the reasons barley is the preferred grain for malting. After settling, the wort is pumped back into the tank, recycled until it is clear, and then pumped off.

The spent grain in the mat is mixed with hot water and allowed to settle; then the residual soluble material is pumped off. This process is called *sparging*. The remaining spent grain is recovered, dried, and used mainly as an animal feed. Although the starch has been removed, the brewer's spent grains are high in protein and crude fiber.

WORT BOILING AND COOLING

The wort is boiled for 1.5–2.0 hr. It has a pH of about 5.5, and boiling at atmospheric pressure sterilizes it at that pH. In addition to killing the microorganisms, the boiling coagulates part of the nitrogenous substances (proteins), producing insoluble material. This phenomenon is called the *hot break*. The tannins from the hops interact with the insoluble proteins, as well as those that are soluble, and make them less soluble. As much as possible of the unstable protein (protein that will become insoluble later at a lower temperature) should be removed.

Generally, one half of the hops is added near the start of wort boiling. Many of the flavor components from both the malt and the hops are volatile and thus are lost during the boiling stage. To retain the appropriate amount of hop flavor and aroma, the remaining half of the hops is added near the end of the boiling stage. During wort boiling, the wort becomes darker in color, presumably because of oxidation.

After the boiling is complete, the wort is cooled, generally to about 7°C. Care must be taken not to infect the wort with microorganisms; therefore, the cooling is in closed tanks. Often, cooled and filtered sterile air is pumped through the wort, not only to aid in cooling but also to raise the oxygen content of the wort. As a result of cooling, more protein and protein-tannin complex are precipitated (this is the *cold break*) and must be removed. After being cooled, the wort is filtered and pumped to pitching tanks.

PITCHING AND FERMENTATION

Pitching is the process of adding yeast to the wort, usually about ¾ lb per barrel. The oxygen added during cooling helps to shorten the lag phase of yeast growth. Generally, a head of foam forms after about 48 hr. The foam carries out some of the remaining hop resins and nitrogenous materials. Because this decreases the bitterness of the beer, the head must be carefully removed.

The wort is then pumped to fermentation tanks. These are generally closed tanks, and part of the CO_2 evolved can be collected to be added back later in the process or sold as a by-product. One third of the carbon in the sugar goes to CO_2 and considerable heat is produced, as is shown by the general reaction sequence:

$$C_6H_{12}O_6 \rightarrow 2\ CO_2 + 2\ CH_3CH_2OH + 22\ \text{calories}\ .$$

Because of the heat produced, the tanks are equipped with cooling coils so that a constant temperature can be maintained. The fermentation rate increases during the first 18 hr of fermentation, holds steady for about 72 hr, and then slowly declines. Fermentation is generally complete in seven to nine days. Its progress can be followed with a saccharometer; sugar-water mixtures have a much higher density than do water-alcohol mixtures. Fermentation is complete when the desired sugar content has been reached. Brewmasters are less concerned with the alcohol content of their beer than with how dry or sweet it is.

When fermentation is complete, the beer is cooled to 4°C and carefully pumped off the sediment in the bottom of the fermentation tank. Among other things, the sediment contains the yeast crop (assuming that a bottom yeast was used). The yeast crop is four to five times the amount of yeast originally used. It is washed and either used for the next pitching or sold as a feed or food ingredient. The growth of yeast is a separate phenomenon from fermentation. However, in brewing, both take place.

The changes that take place during fermentation can be summarized as follows. Sugar is converted to alcohol and CO_2. The pH drops from 5.5 to about 4.3. The nitrogen content falls by about one third, chiefly because of its utilization by yeast. The oxygen content decreases to about 0.3%. Essentially all hop resins are lost (the trace left gives the beer its bitterness). The beer becomes lighter in color, partly because of the pH change. The final specific gravity is about 1.010.

STORAGE AND BOTTLING

The beer is pumped out of the fermentation tank, chilled to 0°C, and filtered through diatomaceous earth. It is *lagered* (aged) at 0°C under a counter pressure of 5-6 psi of CO_2. The aging period varies but in the United States is generally three to 12 weeks. During the lagering period, more settling occurs and the taste becomes more mellow. The mellowing results from the formation of esters from the alcohols and acids produced during fermentation. Yeast is always contaminated with bacteria, which produce both lactic and acetic acid. The acids are responsible for most of the pH drop during fermentation. The aroma of the beer becomes more pleasant and the flavor less yeasty.

Beer can develop a haze that appears as it is cooled and disappears as it is warmed. The term for this reversible phenomenon is *chill haze*. The problem has been traced to a reversible loss of solubility of protein-tannin conmplexes. To avoid the problem (most people like their light-colored beer to be bright and clear), a number of chill-proofs are used. These include tannic acid, proteolytic enzymes, and bentonite (an absorbent clay). The chill-proofs appear to work by adsorbing the protein-tannin complexes. In addition, the beer also develops a greater resistance to haze formation.

After aging, the beer is given a final filtration through cellulose fibers. This is more an absorption than a sieving. It is then ready for bottling. Bottling or canning is done cold and against a counter pressure of carbon dioxide. This avoids the loss of CO_2 and also protects against contamination by organisms or oxygen. Beer is bottled in brown bottles to protect it from light that would alter its flavor.

Beer that is bottled or canned to be stored at room temperature must be pasteurized. This is usually done before it is bottled. High-temperature short-time treatments are generally used. The treatment alters the flavor slightly; thus, bottled beer tastes different from draft (keg) beer. Because draft beer is not pasteurized, it must be kept at low temperature. An alternate method of sterilizing beer is filtering it through micropore filters. This process filters out all bacteria and gives a sterile product that can be canned or bottled. Thus, it is now possible to produce what appears to be a contradiction in terms, bottled draft beer.

Recently, many beers that contain fewer calories (light beers) have come on the U.S. market. A number of approaches can be used to produce a light beer. The simplest is to dilute just by adding water. Unfortunately, or perhaps fortunately, this also has a negative effect on the flavor. Most of the techniques used today center on removing the carbohydrate stubs left in the mash after the α- and β-amylases from the malt have degraded the starch (Fig. 8). These stubs are not fermented by yeast and thus remain in the beer and contribute calories. They are glucose molecules linked α-1,6 and four to eight other molecules linked α-1,4 near the 1,6 bond. If these stubs were reduced to maltose or glucose, the yeast could use them and the calorie content would be less. This can be accomplished by using glucoamylase in the mashing step; green malt, which contains an α-1,6 glucosidase; or an adjunct that contains no starch or α-1,6 bonds, such as a corn syrup with a high dextrose equivalent. The stubs do contribute to the beer's body or mouthfeel. Lack of them has resulted in some people's characterization of light beers as the equivalent of cold dish water. Nevertheless, they remain popular.

Recently there has been a growing market for nonalcoholic beers. In most cases, these are produced by removing the alcohol from beer after it is brewed. The most popular removal method is reverse osmosis.

Distilled Products

Ethyl alcohol for whiskey is obtained by fermenting grain mashes. Industrial alcohol (for fuel or as a chemical feedstock) can also be

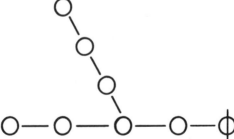

Fig. 8. Stub left after α- and β-amylase attack on starch. ⏀, reducing glucose residue.

obtained from grain but is often produced from cheaper carbohydrates or from petroleum feedstocks.

To produce whiskey, the corn, rye, barley, or other grain is ground, mixed with water, and cooked severely under pressure to gelatinize the starch and sterilize the mash. After cooling, barley malt is added as an enzyme source to convert the starch to sugar. After mashing (which is similar to the process used in brewing), the entire mash is pumped to fermenters and yeast is added. To make a sour mash whiskey, the mash is inoculated with bacteria (either from a pure culture or from a previous mash). Fermentation generally lasts about 72 hr, after which the mash is distilled with no separation of liquid and solids. Thus, the spent grain and yeast are heated during the distilling. The still is usually cut (i.e., a fraction is taken off) at about 160 proof. The term *proof* denotes twice the mixture's alcohol content by volume. Thus, 100 proof is 50% alcohol, the minimum mixture of alcohol and water that will burn. In the old days, igniting the whisky was the "proof" that was often demanded to show that the whiskey was not cut too far.

Distilling to higher proofs (maximum 190 proof) gives a product with no flavor because lower volatile flavor and aroma compounds are lost from the distillate. The whiskey can be diluted to the desired proof. Most whiskey is aged in charred barrels to produce a milder and smoother flavor (charcoal removes aromatic compounds).

In general, the conditions for whiskey production are less tightly controlled than those for brewing. The flavor of the product is determined by the grain used, the inoculum, method and degree of distillation, the conditions of aging, and, if it is a blended whiskey, the skill and taste of the blender.

REVIEW QUESTIONS

1. Why is barley the predominant cereal malted?
2. What is malting?
3. Define dormancy and how it relates to malting.
4. What is after-ripening and how does it relate to malting?
5. Why is barley separated into different classes before malting?
6. What is steeping and what should be accomplished during steeping?
7. What are understeeping and oversteeping?

8. What can be done to prevent oversteeping?
9. What are the normal conditions of germination?
10. What is kilning?
11. During the malting process, what is the normal loss in weight?
12. What is the role of malt in brewing?
13. What adjuncts are used in brewing?
14. What are hops and what do they contribute to brewing?
15. What yeast is used in brewing?
16. What is flocculence and why is it important to the brewmaster?
17. What is accomplished in the adjunct cooker?
18. What is accomplished in the masher?
19. What is wort?
20. What is pitching?
21. Which is more important to the brewmaster, the alcohol or sugar content of the beer?
22. What are the products of brewing?
23. Why is most beer bottled in brown bottles?
24. What is draft beer and how can it be bottled or canned?
25. How can a light beer be produced?
26. What is the major difference between the process of brewing and the production of whiskey?
27. What factors control the flavor of whiskey?

SUGGESTED READING

BRIGGS, D. E., HOUGH, J. S., STEVENS, R., and YOUNG, T. W. 1981. Malting and Brewing Science, Vol. 1. Malt and Sweet Wort, 2nd ed. Chapman and Hall, London.

MEREDITH, P., and POMERANZ, Y. 1985. Sprouted grain. Pages 239-320 in: Advances in Cereal Science and Technology, Vol. 7. Y. Pomeranz, ed. Am. Assoc. Cereal Chem., St. Paul, MN.

OHLMEYER, D. W., and MATZ, S. A. 1970. Brewing. Pages 173-220 in: Cereal Technology. S. A. Matz, ed. Avi Publishing Co., Westport, CT.

PALMER, G. H., and BATHGATE, G. N. 1976. Malting and brewing. Pages 237-324 in: Advances in Cereal Science and Technology, Vol. 1. Y. Pomeranz, ed. Am. Assoc. Cereal Chem., St. Paul, MN.

POLLOCK, J. R. A. 1987. Brewing Science. Vols. 1-3. Academic Press, London.

WITT, P. R., Jr. 1970. Malting. Pages 129-172 in: Cereal Technology. S. A. Matz, ed. Avi Publishing Co., Westport, CT.

YOSHIZAWA, K., and SOHTARO, K. 1985. Rice in brewing. Pages 619-645 in Rice: Chemistry and Technology, 2nd ed. B. O. Juliano, ed. Am. Assoc. Cereal Chem., St. Paul, MN.

Gluten Proteins

CHAPTER 10

The *gluten proteins* are the storage proteins of wheat. They are easy to isolate in relatively pure form because they are insoluble in water. The starch and the water solubles can be removed from the gluten by gently working a dough under a small stream of water. After the washing, a rubbery ball of gluten is left. Gluten was first isolated by this method by Beccari in Italy in 1728. It was the first protein isolated from plant material; before that time, proteins were thought to come only from animal sources. As isolated from flour, gluten contains (on a dry basis) about 80% protein and 8% lipids, with the remainder being ash and carbohydrate.

When wheat flour is mixed with water, a cohesive, *viscoelastic dough* is formed. No other cereal flour forms a dough with a similar viscoelastic character. It is generally believed that the gluten proteins are responsible not only for this cohesive, viscoelastic property of wheat flour dough but also for the dough's ability to retain gas during fermentation and, at least partly, for the setting of the dough during baking.

Why gluten proteins interact with themselves to form a strong dough is still largely unknown. However, several factors may be related to the dough-forming ability of gluten. The amino acid composition in Table I shows that the gluten proteins are very high in glutamic acid (about 35% of the total protein). Stated another way, one in every three amino acids in gluten is glutamic acid. The glutamic acid residue occurs in the protein as its amide, *glutamine*, rather than as the free acid. Evidence for this is the high ammonia nitrogen found after acid hydrolysis and the fact that gluten proteins do not migrate under electrophoresis in alkaline buffers. Failure of the proteins to migrate over an alkaline range of pH values is taken as evidence that practically no negative charges exist on the gluten proteins in alkaline media.

The gluten proteins are also notably low in the basic amino acids. The low level of lysine is well known. Thus, the gluten proteins have essentially no potential negative charges and only low levels of potential positive charges. The conclusion from these facts must be that the gluten proteins have a low charge density. That low level of charges means that the repulsion forces within the protein are low and, thus, the protein chains can interact with each other quite readily, a condition that appears to be necessary for dough formation.

The next most notable point about gluten's amino acid composition is the high level of *proline*, about 14% of the protein or about one in seven residues. Because proline's amino group is involved in a ring structure, peptide bonds of proline are not flexible. Thus, there is a rigid kink in the protein chain wherever proline occurs, and the protein cannot readily form into an α-helix. Measurements of helical structure in gluten have generally given low values. This does not necessarily mean that the gluten proteins have no ordered structure, only that it is not the α-helix with which we are familiar.

Interestingly, differential scanning calorimetry thermograms for wheat gluten fail to show a definite *denaturation* peak (Fig. 1). This

TABLE I
Amino Acid Composition (mol/10^5 g of protein)
of Gluten, Gliadin, and Glutenin[a]

Amino Acid	Gluten	Gliadin	Glutenin
Arginine	20	15	20
Histidine	15	15	13
Lysine	9	5	13
Threonine	21	18	26
Serine	40	38	50
Aspartic acid	22	20	23
Glutamic acid	290	317	278
Glycine	47	25	78
Alanine	30	25	34
Valine	45	43	41
Leucine	59	62	57
Isoleucine	33	37	28
Proline	137	148	114
Tyrosine	20	16	25
Phenylalanine	32	38	27
Tryptophan	6	5	8
Cystine/2	14	10	10
Methionine	12	12	12
Ammonia	298	301	240

[a] Source: D. D. Kasarda, C. C. Nimmo, and G. O. Kohler. 1971. Pages 227-299 in: Wheat Chemistry and Technology, 2nd ed. Y. Pomeranz, ed. Am. Assoc. Cereal Chem., St. Paul, MN. Data from Y. V. Wu and R. J. Dimler, Arch. Biochem. Biophys. 102:203 (1963) and 103:310 (1963). Used by permission.

could be interpreted to show that there is no ordered structure in the gluten proteins. Clearly, much more work is required before we understand the structure of gluten proteins.

The remaining amino acids in gluten are not very unusual; they consist of reasonably high levels of amino acids with hydrophobic side chains and relatively low amounts of sulfur-containing amino acids. In general, the amino acid composition of gluten shows that about half the protein is made up of two amino acids (glutamine and proline). The charge density of the protein is extremely low, with low levels of basic amino acids and with practically all the acid groups occurring as their amides. The high amide content and the low charge on the protein would suggest substantial hydrogen bonding in the system. Evidence for the importance of hydrogen bonding is shown by mixing flour with D_2O instead of H_2O (Fig. 2). D_2O is known to have stronger

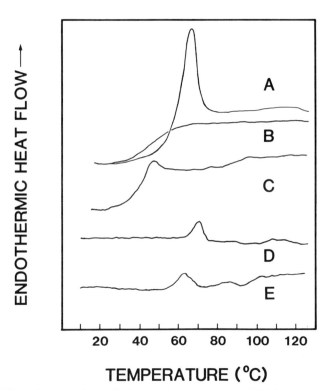

Fig. 1. Differential scanning calorimeter thermograms at 10°C/min. A, commercial wheat gluten (11.3% moisture); B, sample A cooled and reheated; C, hand-washed gluten (11.4% moisture); D, hand-washed gluten, excess solvent (30% glycerol); E, hand-washed gluten, excess water.

hydrogen (deuterium) bonding. As shown by the width of the mixograms, the dough produced is much stronger. If a mixogram is produced with a reagent that breaks hydrogen bonds, such as urea, the dough is much weaker (Fig. 3).

The amino acid composition of gluten proteins also shows that about 35% of the total amino acids have hydrophobic side chains. It has been suggested that when the level of hydrophobic side chains is greater than 28%, the apolar residues are not accommodated in a hydrophobic core of the protein. Thus, the amount of surface hydrophobicity on the protein increases, encouraging hydrophobic interactions. Therefore, it is believed that hydrophobic interactions between gluten proteins play an important role in stabilizing gluten structure and in the rheological and baking properties of flour.

Further evidence of the gluten proteins' hydrophobic nature is their effectiveness in binding lipids. Wheat flour contains about 0.8% lipids extractable with petroleum ether. However, after the flour is wetted and mixed into a dough, only about 0.3% is extractable from the lyophilized dough. Presumably the remaining 0.5% was hydrophobically

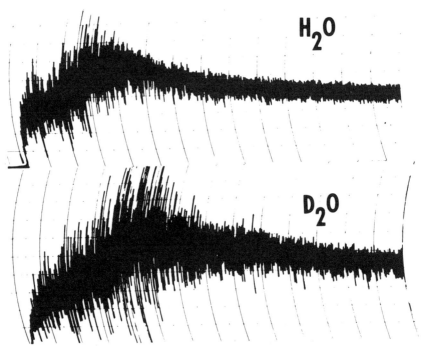

Fig. 2. Mixograms of wheat flour mixed with H_2O and D_2O.

bound to the proteins. Gluten also binds lipids added to flour during mixing.

Physical Properties of Gluten

The gluten complex is composed of two main groups of proteins: *gliadin* (a prolamin) and *glutenin* (a glutelin). The two protein groups can be conveniently separated by solubilizing the gluten in dilute acid, adding ethyl alcohol to make the solution 70% alcohol, then adding sufficient base to neutralize the acid. After standing overnight at 4°C, the glutenins precipitate, leaving the gliadins in solution.

The gliadins are a large group of proteins with similar properties. They have an average molecular weight of about 40,000, are single-chained, and are extremely sticky when hydrated (Fig. 4). They have little or no resistance to extension, and appear to be responsible for the dough's cohesiveness.

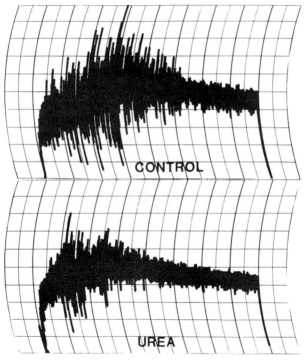

Fig. 3. Mixograms for a control flour and the same flour containing 1.5% urea (based on flour).

The glutenin proteins also appear to be a heterogeneous group of proteins. They are multichained and vary in molecular weight from about 100,000 to several million. Their average molecular weight is about 3 million. Physically, the protein is resilient and rubbery but prone to rupture (Fig. 4). The glutenin apparently gives dough its property of resistance to extension.

Starch gel or polyacrylamide gel electrophoresis has been used to characterize the two groups of proteins. The gliadins migrate to give a large number of bands (Fig. 5). The glutenins, in general, do not migrate into starch gels because they are too large to enter the gel pores. Instead, the glutenin packs at the surface of the gel and forms streaks in the gel. In electrophoresis of solutions (not gels), most of the glutenin migrates as one band with a mobility similar to that of the fastest-moving gliadin band. Essentially all electrophoretic patterns of gluten have faster-moving bands that are albumin and globulin proteins; these are considered to be contaminants and not part of the gluten complex. However, there is no direct evidence that some of those proteins are not involved in the gluten complex. The heterogeneity of gliadin can be illustrated better by a combination of isoelectric focusing and polyacrylamide electrophoresis (Fig. 6). As many as 50 gliadin proteins have been identified by that technique.

Another useful technique, particularly for the glutenin proteins, is sodium dodecyl sulfate electrophoresis after the disulfide bonds have been reduced with a reagent such as mercaptoethanol (Fig. 7). After the glutenin is reduced, it can migrate into the gel. The degree of

Fig. 4. Physical properties of gluten (left) and its components gliadin (center) and glutenin (right).

migration is related to each fraction's molecular weight. The molecular weights of the glutenin subunits vary from less than 16,000 to about 133,000.

Thermal Properties of Gluten

A number of workers have shown that wheat gluten-water mixtures, when heated in a differential scanning calorimeter, do not give a denaturation peak (trace E, Fig. 1). The small peaks that are seen apparently are caused by contaminating starch granules or soluble proteins. The lack of a denaturation peak would suggest that the gluten is in a random configuration. If the proteins are in random configuration, they certainly cannot go to another random structure. Random is random. On the other hand, if the proteins have a three-dimensional

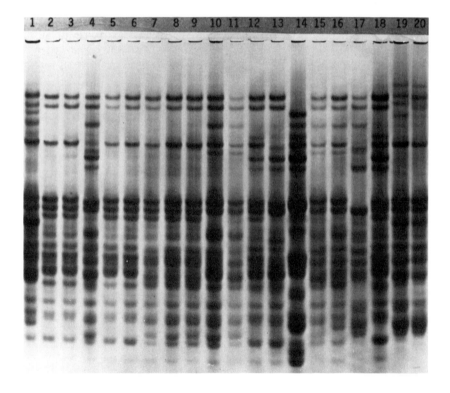

Fig. 5. Polyacrylamide gel electrophoresis of the gliadin proteins from a number of wheat cultivars.

structure that goes to random upon heating, then a denaturation peak in the differential scanning calorimeter should result.

It has been shown by *circular dichroism* that gluten contains β-turns when dissolved in certain solvents. While on the surface the two reports contradict each other, it appears possible that both studies could be correct because of the widely different test conditions.

If gluten is heated in the differential scanning calorimeter under low moisture conditions, a *glass transition* is clearly shown (see the change in the baseline—heat capacity—in traces A–C in Fig. 1). Glass transition is a large change in physical properties occurring over a relatively narrow temperature range; it is discussed in Chapter 14. As would be expected, the glass transition temperature (T_g) decreases rapidly as water is added to gluten (Fig. 8). Examination of the curve shows that gluten's T_g approaches room temperature at about 16% moisture. Thus, these are the temperature and moisture conditions that produce a leathery protein; if either moisture or temperature is increased, the protein become rubbery. Therefore, we do not need any specific configuration of the protein to produce the partially elastic properties of gluten, but only the correct moisture and temperature.

Genetics of Gluten

The number of proteins found in wheat cultivars, as well as the amount of each protein, appears to be mainly determined by the *genotype*.

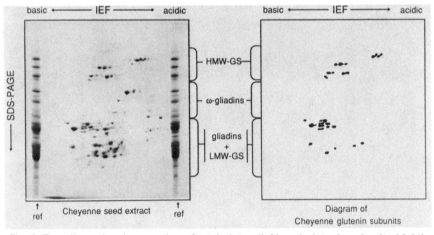

Fig. 6. Two-dimensional separation of total gluten (left) and glutenin subunits (right) by isoelectric focusing (IEF) and sodium dodecyl sulfate-polyacrylamide gel electrophoresis (SDS-PAGE). GS, glutenin subunits; HMW, high-molecular-weight; LMW, low-molecular-weight.

Stated in different words, it is the genetics of the wheat variety that determines which proteins are present and the relative amount of each protein. The overall amount of protein in a sample is affected strongly by the environment and to a lesser extent by the genotype. Interestingly, the environment appears to have relatively little effect on the number of proteins or the relative amount of protein in a wheat sample. This fact has allowed methods to be developed to identify wheat cultivars by their electrophoretic patterns. Relatively small amounts of protein are needed; thus, the analysis can be performed on a single kernel of wheat. Actually, the germ can be excised and saved for planting, while the remainder of the kernel is ground and extracted.

If one is attempting to identify an unknown wheat variety, it is often an advantage to analyze a number of single kernels from the sample rather than to grind a lot and extract that sample. If the sample is a mixture, that fact becomes apparent when single kernels are analyzed. Commercial cultivars are seldom pure—they are often a mixture of related genotypes. Under different environments, the ratio of the different genotypes may vary. The ratio of different proteins

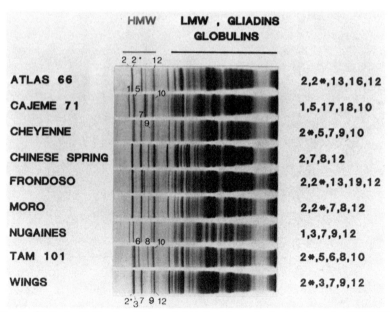

Fig. 7. Sodium dodecyl sulfate-polyacrylamide gel electrophoresis (SDS-PAGE) analysis of nine wheat cultivars. The numbers at the right give the high-molecular-weight (HMW) subunits numbered according to the nomenclature of P. I. Payne and G. J. Lawrence.

in a cultivar has also been shown to vary under very severe conditions, for example, when sulfur in the soil is severely deficient.

Synthesis of Gluten

The development of the wheat endosperm is extremely complex. Gluten has been reported in the endosperm as early as six days after *anthesis* (flowering). *Protein bodies*, the site of gluten deposition in the endosperm, have been reported as soon as 12 days after anthesis. The protein bodies increase in size, up to 20 μm in diameter, as the endosperm develops. When the grain is mature and the kernel starts to desiccate, the protein bodies are distorted by the starch granules and completely lose their identity in the mature seed. Protein bodies that have been

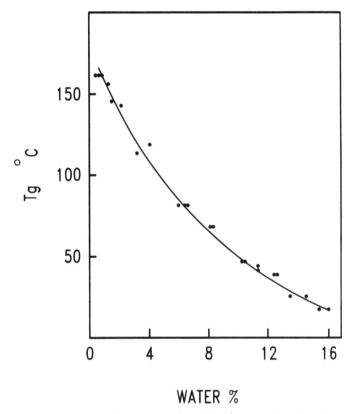

Fig. 8. Change in glass transition temperature (T_g) as a function of moisture content of hand-washed and commercial glutens.

isolated from immature seeds were shown to contain gluten, both gliadin and glutenin, and essentially no albumins and globulins.

Common bread wheat is a *hexaploid*—that is, it contains three genomes, A, B, and D. On the other hand, durum wheat is a *tetraploid* and contains only the A and B genomes. Scientists have found the location on the chromosome of some of the genes for specific proteins and polypeptides.

Large Gluten Aggregates

It appears clear that gluten controls the baking quality of bread wheat flours. It is also probably true that it controls much of the baking quality of a number of other products. Early in this century, T. B. Osborne unsuccessfully tried to relate the ratio between gliadin and glutenin to the baking quality of bread. More recently, a number of studies have shown that the strength of a dough is related to the amount of less-soluble protein. The less-soluble protein is the glutenin protein and most likely the *high-molecular-weight* (HMW) glutenin subunits. This fact has led to a number of tests for protein strength. However, it should be kept in mind that gluten strength and loaf volume potential (the ability to retain gas) are not necessarily closely related. Clearly, dough strength and mixing time are both related to the amount and type of HMW proteins.

At the same time, it is also clear that the gliadin proteins provide the dough with extensibility. The action of the gliadin proteins appears to be associated, at least partially, with the sulfur-containing gliadin proteins. The ω-gliadins that contain no sulfur are not very effective at increasing extensibility. Presumably, the sulfur is important because of its involvement in disulfide interaction reactions. The importance of those interactions is also shown by adding reagents that block thiol groups to dough. The dough is strong in the presence of these reagents.

Individual Proteins or Polypeptides

A tremendous amount of work has been done attempting to determine what individual protein or small group of proteins is most effective in determining baking quality. To date, these studies have been essentially unsuccessful. The main success has been with durum wheats, in which the presence of a certain gliadin band (45) gives good pasta and the presence of gliadin band 42 is related to poor pasta. Although the association is very good, it now appears that the gliadin bands are genetic markers rather than causative agents. In the hexaploid wheats, such associations appear to be very complex. Much work has concentrated on the HMW subunits of glutenin. There appears to be no question that if these subunits are missing, the dough is very weak. It is also true that certain subunits have been linked to better strength (i.e., 5 and 10 are better than 2 and 12, using the nomenclature of Payne and Lawrence[1]). But as pointed out above, strength and loaf volume potential may not be strongly associated. Clearly, doughs can be too strong. Such doughs are termed "bucky" doughs in the bakery and produce loaves with poorer loaf volume, poor crumb grain, and very poor symmetry. In some countries (Argentina, for example), the wheats that produce bucky doughs are a major quality problem.

In taking an overall view of this work, one must consider that if the total amount of protein is highly correlated with loaf volume potential, then quality is probably not controlled by any one or even a small number of proteins or peptides. If there *were* one "key" protein, we must assume that it would be highly correlated with the total protein content of the sample.

Commercial Isolation of Gluten

Worldwide, the commercial production of wheat gluten is of the order of 300,000 metric tons per year. The largest production is in the European Community (EC), followed by the United States, Australia, and Canada. The United States is by far the largest importer of wheat gluten. It appears strange that the United States, with its large excess wheat production, would be such a large importer of gluten. The reason appears to be that for each kilogram of gluten produced there is somewhere near 7.0 kg of starch produced. However, of this 7.0 kg of starch, only about 5.5 kg is prime, or A, starch. The remaining 1.5 kg is tailings, or B, starch. The value of the B starch is much less than that of the A starch. It is used as a feed material for livestock or as the starting material for ethyl alcohol production.

The market for the gluten is generally quite good. However, sale of the wheat starch meets with a number of problems. First, there are relatively few unique uses of wheat starch that make it the desired product. Most wheat starch must be sold in direct competition with corn starch. In the manufacturer of corn starch there is a much higher yield of A starch. Also, the starch market in the United States has traditionally been corn starch, and the users of starch are familiar with its properties. They see little reason to switch to wheat starch

[1]Payne, P. I., and Lawrence, G. J. 1983. Cereal Research Communications 11(1):29-35.

and learn to deal with a new set of properties. Thus, selling wheat starch is not so easy.

In the EC and in Australia, relatively little or no corn is grown. The corn that *is* grown is used primarily as animal feed and not for starch manufacturer; the traditional starch of commerce is wheat starch. Because there is a market for wheat starch, wheat gluten can be produced at a competitive price. This makes it more competitive in the United States and leads to the U.S. importation of large amounts of gluten.

Uses of Wheat Gluten

The largest use of wheat gluten is by the baking industry. In the EC, it is a common practice to add wheat gluten at the mill to increase the protein content of the flours and improve their strength. This is not commonly done in the United States. Most of the U.S. usage of gluten is at the bakery. Many formulas for specialty breads, those containing various fiber sources, for example, contain relatively large amounts of gluten. Sometime the gluten use is amazingly high, i.e., 20% of the flour weight. The gluten is added to strengthen the dough and to produce a product with good texture and volume.

A second major use of gluten at the bakery is in the production of hamburger and hot dough buns. Gluten serves two purposes. First, it produces a bun with a strong *hinge* (when buns are sliced, a small part is not cut so that the two halves of the bun remain attached; the hinge is that uncut part). Second, it is the fad in fast food restaurants to serve hamburger not only hot but also with a sauce. The sauce, combined with the steam, weakens the buns and allows the sauce to come into contact with the customer's hand. Addition of gluten helps the bun to withstand this abuse.

A third use is to allow the bakery to inventory only one flour. Bakeries commonly used to have a number of flours in inventory at any time. If a strong, higher-protein flour was needed for hearth bread, it was available. Now it is more economical to inventory only one flour (white pan bread flour) and use gluten in the formula if a stronger flour is desired.

Other areas where gluten is used include breakfast cereals (e.g., Special K), where it is used to increase the protein content and also to adhere to other ingredients. In many pet foods, the gluten acts as a *meat analog*. Gluten does a good job of aligning to give the appearance of meat texture. The pets probably do not care whether the product looks like meat, but then the pets do not buy the pet food. Gluten is also used as a binder and texturizing agent in many sausage-type products.

Nutritional Quality

Wheat gluten is low in the basic amino acids. The most severe limit is lysine; when used in an animal diet, the first limiting amino acid is lysine. If protein is limited in the diet, addition of other proteins with high lysine contents or addition of the free amino acid will greatly improve the protein utilization. However, this is not true if the level of protein is not limiting. The average U.S. person eats many times his or her daily requirement for protein, and, therefore, the amino acid balance and the lysine content of the protein are of little or no importance. Feeding trials with young males (college students) have shown that when bread supplied 90% of the protein in the diet, the subject stayed in positive nitrogen balance.

REVIEW QUESTIONS

1. What factors about the amino acid composition of wheat gluten probably make the proteins interact to form a cohesive gluten?
2. What evidence suggests that glutamine, rather than glutamic acid, occurs naturally in gluten?
3. What are the two most predominant amino acids in gluten?
4. What evidence suggests that hydrogen bonds are important in gluten structure?
5. What evidence suggests that hydrophobic bonds are important in gluten structure?
6. What is gliadin and what are its properties?
7. What is glutenin and what are its properties?
8. What is a good reagent to reduce glutenin to its subunits?
9. Why is it surprising that gluten gives no denaturation peak in a differential scanning calorimeter?
10. At about what moisture content does gluten give a glass transition at room temperature?
11. What is necessary for a polymer to have elastic properties?

12. What determines the number of proteins found in a cultivar?
13. Why is it an advantage to measure individual kernels if you are attempting to identify cultivars by their electrophoretic patterns?
14. Bread wheat is a hexaploid and durum is a tetraploid. What does this mean?
15. Why is the United States a large importer of gluten?
16. What are a number of uses of wheat gluten?

SUGGESTED READING

BUSHUK, W., and TKACHUK, R., eds. 1991. Gluten Proteins 1990. Am. Assoc. Cereal Chem., St Paul, MN.

GARCIA-OLMEDO, F., CARBONERO, P., and JONES, B. L. 1982. Chromosomal locations of genes that control wheat endosperm proteins. Pages 1-47 in: Advances in Cereal Science and Technology, Vol. 5. Y. Pomeranz, ed. Am. Assoc. Cereal Chem., St. Paul, MN.

GRAVELAND, A., and MOONEN, J. H. E., eds. 1984. Gluten Proteins: Proceedings of the 2nd International Workshop on Gluten Proteins. TNO, Wageningen, Netherlands.

LASZTITY, R., and BEKES, F., eds. 1987. Gluten Proteins: Proceedings of the 3rd International Workshop on Gluten Proteins. World Scientific, Singapore.

WRIGLEY, C. W., and BIETZ, J. A. 1988. Proteins and amino acids. Pages 159-275 in: Wheat: Chemistry and Technology, 3rd ed. Vol. 1. Y. Pomeranz ed. Am. Assoc. Cereal Chem., St. Paul, MN.

WRIGLEY, C. W., AUTRAN, J. C., and BUSHUK, W. 1982. Identification of cereal varieties by gel electrophoresis of the grain proteins. Pages 211-259 in: Advances in Cereal Science and Technology, Vol. 5. Y. Pomeranz ed. Am. Assoc. Cereal Chem., St. Paul, MN.

Rheology of Doughs and Batters

CHAPTER 11

Doughs and batters are complex mixtures of many ingredients. Their complexity is not restricted to their composition in a chemical sense but also includes their physical properties. If you watch an "old-time" baker work with dough, you will see him or her touch the dough, pull on it, squeeze it, and perhaps roll it between his or her hands before pronouncing that it will or will not run on the line. Of course, the baker is making rheological measurements on this dough. Asking the baker the results of the rheological measurements will, at best, bring a puzzled look. An experienced baker can tell much about a dough or batter by noting its appearance and other physical properties.

The problem comes when the information is transferred to another baker or, worse yet, to a new college graduate who has recently joined the company. The baker's instructions might be "mix the dough a little short; keep it slightly slack; use a flour that won't give a bucky dough; and be sure to skin the dough before make-up." Only experience allows one to translate those instructions.

Attempts to transfer the information in a more rigorous manner have resulted in many rheological devices. Some of these instruments were designed to determine the amount of mixing a dough requires or the amount of water that should be added to the flour. They were also found useful to characterize the various flours or other ingredients being used. Some of the instruments were designed for dough, but many were adaptations of instrument designed to study other materials, e.g., plastics.

Rheology

Rheology is the study of how materials deform, flow, or fail when force is applied. The rheological properties of some materials can be described by a single value. For example, the flow of water is defined by its viscosity. The deformability of a steel spring is defined by Hooke's constant (a modulus of elasticity). However, most materials, and certainly doughs and batters, are not that simple in their properties or behaviors but show a more complex rheological behavior.

If a material's viscosity is constant regardless of shear rate (the rate of stirring or flowing through a pipe), the material is said to show *Newtonian* or *ideal viscosity*. Its behavior then can be defined by a single viscosity value. However, in many systems, including most flour-water systems, the viscosity changes (decreases) as the shear rate is increased. Thus, the system shows more complicated *non-Newtonian* behavior, and we cannot define the system by a single viscosity value but must give a viscosity at each shear rate. In addition, viscosity can also be affected by the time involved in making the measurement.

There are many different kinds of moduli. In general, *modulus* refers to the stiffness of the material and is a proportionality constant relating *stress* to *strain*. In lay terms, it tells how much force is required to produce a specific *deformation* of the material under test.

In the cereal chemistry literature, we often see measurements taken with a farinograph or mixograph referred to as rheological measurements. Certainly, these instruments measure how doughs deform and flow. Therefore, they clearly fit our definition of rheological study. The problem with the use of these instrument for rheological studies is that we cannot define the stress on the sample at any moment of time during the test. For example, in a mixograph bowl, only a small part of the dough is in contact with a pin at any given time, and the shape of the sample (dough) changes in very complicated and unpredictable ways. Thus, it is impossible to determine the stress on the dough, as we do not know the geometry of our test piece. As a result, the measurements made using a mixograph are valid only for the mixograph, and measurements made using the farinograph are relevant only to the farinograph.

This is not to say that the above instruments are not useful. They have stood the test of time and can give much useful information. They are particularly useful when used to characterize or "fingerprint" a flour. We often want to know whether the mixing properties of the flour we are using today are similar to or different from those of the flour we used yesterday. The mixograph or farinograph can easily and rapidly answer that question for us. This chapter does not concentrate on those types of rheological measurements but instead looks at more fundamental rheological measurements.

The reason for this emphasis is twofold. First, from fundamental measurements we can learn more about our dough or batter. We can see the effect of various interactions and how the properties of the dough or batter change as a function of time or temperature. In addition, the measurements can often be made on the complete dough or batter system so the results are relatively easy to interpret. The second reason is that, with the advent of minicomputers and their related equipment, it has now become much easier to obtain good rheological data.

Dynamic Rheological Measurements

Dynamic rheometers have been used quite successfully in the plastics field and have relatively recently been used in the measurement of dough rheological properties. The type and geometry of a dynamic rheometer can vary rather widely, but the basic measuring principles are the same regardless of the testing geometry. Those basic principles are illustrated in a parallel plate testing mode, in which one plate is fixed to a force transducer and the other plate is oscillated in a sinusoidal motion. This is illustrated schematically in Fig. 1. As the top plate is moved, stress is applied to the dough sample between the plates.

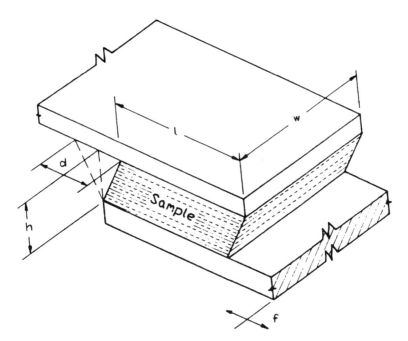

Fig. 1. Parallel plate geometry for dynamic testing. d = deformation, f = force transmitted through sample, h = sample thickness, l = length of sample, w = width of sample.

Stress is defined as a force acting over a unit area, often in the units called Pascals (N/m²). It is usually referred to by the symbol σ. When sufficient force is applied to the dough, the dough deforms, and the deformation produces a strain in the dough. Here the strain is the amount of deformation divided by the height or thickness of the dough. Strain is usually referred to by the symbol γ and is measured as a percentage. For example, if the deformation is 1% of the thickness of the dough, the dough is under a 1% strain.

In Fig. 1, the top plate, which is in contact with the sample (it must not slip), is caused to oscillate sinusoidally at some frequency (ω, in radians per second) and with an amplitude (d) measured in millimeters. The bottom plate remains stationary and is attached to a force transducer. If no slippage occurs at either plate, a deformation gradient is created across the thickness of the sample (h). The force transducer thus measures the force (f) transmitted through the sample in Newtons. The force is distributed over the sample area (1 × w) and is uniform over the sample thickness. The output of such a dynamic rheometer is given in Fig. 2. A deviation from sinusoidal behavior would indicate that slippage is occurring. The τ (tau) curve is the output of the force

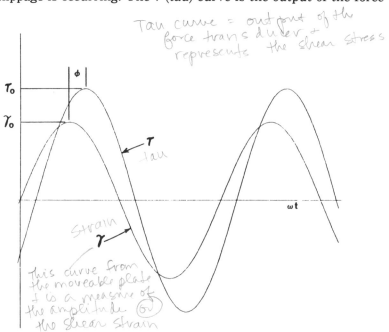

Fig. 2. Sinusoidal signal output from the force and deformation transducers. τ_o = shear stress amplitude, γ_o = shear strain amplitude, ϕ = phase angle, ωt = radians/sec.

transducer and represents the shear stress. The γ curve is from the moveable plate and is a measure of the amplitude or the shear strain. If the sample were completely elastic, the two curves would peak together, and if the sample were completely viscous, the two curves would be out of phase by 90°. The *phase angle* phi (ϕ), in radians, gives a measure of how much the system is out of phase. The *complex modulus* (G^*) is equal to τ_o/γ_o. As the name implies, the complex modulus is made up of the *storage modulus* (G') and the *loss modulus* (G''). The storage modulus and the loss modulus are given by the following equations:

$$G' = (\tau_o/\gamma_o) \times \cos \phi$$

$$G'' = (\tau_o/\gamma_o) \times \sin \phi$$

In lay terms, G' is the part of the energy that is stored during a cycle, and G'' is the part that is lost. Another term that is often used is tan δ or tan ϕ, which is the ratio G''/G'. This is a simple index of the relative elastic or viscous nature of the material under test.

Other Rheological Measurements

EXTENSIGRAPH

The extensigraph has been widely used in both quality control and research laboratories for studying flour quality and the effect of certain additives in bread baking. This load-extension instrument was designed in the 1930s to provide empirical measures of stress-strain relationships in doughs. After being mixed, dough is scaled, molded into cylindrical shapes, and clamped at both ends in a special cradle. After a rest period, the dough is stretched through the its midpoint with a hook that moves at a constant rate of speed until the dough ruptures. The result is a load (resistance) vs. extension curve called an extensigram (Fig. 3). Interpreting extensigraph measurements in terms of basic physical or rheological terms is difficult because the dough geometry changes constantly as the test is being performed. However, several useful measurements can be made.

The measurements most commonly obtained from extensigrams are; R_m, maximum resistance (the maximum height of the curve); R_5, the resistance at an extension of 5 cm; and the total curve length in centimeters. The R values are often given in arbitrary units of resistance called Brabender units. Occasionally, one finds the total area under the curve reported in square centimeters. For most practical applications, the curve height and area under the curve are taken as measures of the flour's strength, with larger values indicating greater strength. The overall shape of the curve, or the ratio of curve height (h) to extensibility (e), gives an estimation of the dough's viscoelastic balance. Obviously, long, low curves produce low h/e ratios and a predominance of viscosity over elasticity.

By using these simple relationships, extensigrams can classify flours according to their strength: weak, medium, strong, and very strong (as illustrated in Fig. 4). The instrument is also quite useful in studying anything that alters the strength of the dough. Examples of this are proteolytic enzymes and various oxidizing or reducing agents. The

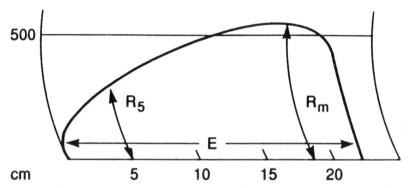

Fig. 3. Extensigram, showing extensibility (E), resistance at a constant extension of 5 cm (R_5), and maximum resistance (R_m).

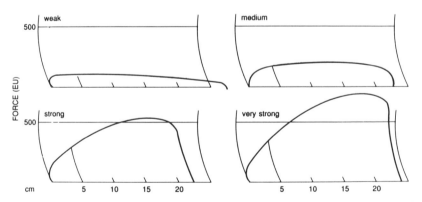

Fig. 4. Extensigrams of flours with weak, medium, strong, and very strong dough properties.

extensigraph has also been used to follow the *structural relaxation* of doughs. This involves measuring the R_5 or R_m as a function of time. This clearly shows the relaxation of dough.

ALVEOGRAPH

The alveograph was invented by Marcel Chopin in the 1920s as an empirical instrument to measure flour "quality." The instrument blows a bubble of dough and measures the pressure during the inflating operation. The dough is mixed, sheeted into a flat piece, and secured in the instrument, and then air pressure is used to blow the bubble. Presumably, the idea behind the test was that blowing a bubble is related to the expansion of bubbles (gas cells) in fermenting dough.

A typical alveogram is given in Fig. 5, and the important measurements that are usually made are identified on the figure. At first glance, the alveogram appears to be a force-time (or load extension) curve similar to that produced by an extensigraph or other similar instrument. However this is not the case. Whereas the extensigraph stretches the dough

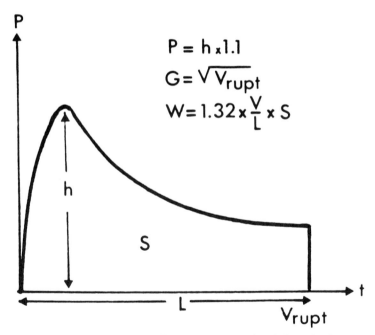

Fig. 5. A representative alveogram. P = overpressure (mm), L = abscissa at rupture (mm), G = swelling index (ml), V = volume of air (ml), W = deformation energy (10^{-4} J), h = maximum height (mm), and S = area under the curve (cm^2).

in a *uniaxial* mode (along a single axis), the alveograph stretches the dough sheet in all directions (Fig. 6). This is *biaxial* stretching, which may have advantages, as it is the type of expansion that actually occurs in fermenting dough.

The commonly measured values from the alveogram are the height of the curve, its length, and the area under the curve. The P value (or overpressure) is the height of the curve multiplied by a constant. The overpressure gives, in millimeters of water, the maximum pressure attained during inflation of the bubble. It is not clear exactly what property the maximum height of the alveogram curve is measuring. Even so, the height of the curve (or overpressure) is widely used in interpretation of alveograms.

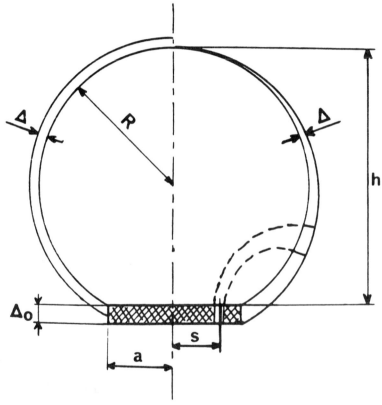

Fig. 6. Geometric characteristics of a dough bubble: constant thickness hypothesis (left), Bloksma's model (right). Δ = bubble thickness, Δ_o = 0.25 cm, radius a = 2.74 cm, R = bubble radius, h = bubble height, s = distance between a dough element and the axis of the disk before inflation.

The average curve length to bubble rupture (L) is clearly a measure of the dough's extensibility. Other important parameters obtained from alveograms are the area under the curve and the deformation energy (W). The W value is calculated as $W = 1.32\ (V/L)(S)$, where V is the volume of air (in milliliters) displaced, L is the length of the curve in millimeters, and S is the area under the curve in square centimeters. W is often considered to be a measure of the flour strength and therefore the most important value derived from the alveogram.

Unlike extensigrams, alveograms have been used to calculate fundamental rheological data concerning wheat flour dough. The calculations are tedious. The reader is referred to the discussion in *The Alveograph Handbook*, which is listed at the end of the chapter, for a more detailed discussion.

LUBRICATED UNIAXIAL COMPRESSION

The rheological properties of fermenting doughs are particularly difficult to determine. In fact, most of the techniques discussed above do not give satisfactory results with fermented doughs. Part of the problem is not only that such doughs contain gas, but that the ratio of gas to dough is always changing during fermentation. The use of a specific relaxation technique called lubricated uniaxial compression appears to be one technique that produces good results from fermented dough.

The test is conveniently run with a universal testing machine. The metal platen is coated with a adhesive-backed Teflon sheet, while the dough sample is coated with a paraffin oil. The Teflon and mineral oil eliminate any friction between the dough sample and the instrument. The sample is compressed at a predetermined rate to a predetermined strain. The crosshead is then stopped, and the stress allowed to decay. The result is a stress relaxation curve (Fig. 7). Biaxial elongational viscosity (n_{be}) can be calculated from the data in the curve.

Doughs

Wheat flour dough is *viscoelastic*, i.e., it exhibits both viscous flow and elastic recovery. Viscous flow means that the material will flow under stress and not recover immediately when the stress is released. When a piece of dough is placed on a flat surface, if the humidity is high enough so that it does not develop a skin, it will flow. The amount of flow depends on the balance of viscous and elastic properties. Good examples of other common viscous materials are honey or motor oil on a cold morning.

The elastic property that we are all familiar with is that exhibited by a rubber band. The common rubber band is truly elastic. By this we mean that when the band is deformed and we remove the force causing the deformation, the band returns to its original size and shape essentially instantaneously. To exhibit these properties, the rubber must have a large molecular weight and be highly cross-linked.

Natural rubber (i.e., that coming directly from the rubber tree) is of large molecular weight but is not cross-linked. In fact, natural rubber is viscoelastic. A ball made from natural rubber, if left overnight, flows to become a flat pool. It can be rounded into a ball again if you desire. Rubber to make rubber bands, and for that matter essentially all rubber products we are familiar with, has been chemically cross-linked by adding sulfur to the rubber and heating the mixture. The process, called vulcanization, was invented by a Mr. Goodyear (you may have heard of him or his blimp).

As mentioned above, dough (like natural rubber) has flow properties. But because dough is not highly cross-linked, as vulcanized rubber is, it is not truly elastic. If we stretch a piece of dough rapidly and release the force immediately, it will partially recover its original shape. The best example of this is rolling out a pizza dough. If we have it rolled as we like and then stop for a rest or a drink of our beer, we

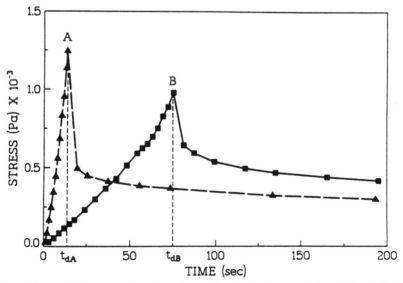

Fig. 7. Stress (Pa) versus time (sec) for a chemically leavened dough in lubricated uniaxial compression. After deformation to 70% in time T_d, deformation is held constant and stress relaxation occurs. Deformation rates: A, 5 cm/min; B, 1 cm/min.

find the dough has recoiled elastically to a new (smaller) size. Another way of illustrating the difference between dough and vulcanized rubber is by stretching a piece of dough and holding it in the extended position for some period of time. When we release the force on the dough, it recoils little, if at all; the stress has relaxed during the rest time, and the dough no longer has elastic properties.

The explanation for the difference in behavior of vulcanized rubber and dough is that the large polymers that comprise the rubber are cross-linked by covalent bonds. In dough, the cross-links between molecules are secondary (noncovalent) bonds that are constantly breaking and reforming. That breaking and reforming allow the high-molecular-weight polymers of gluten to relax when stressed and thus lose their strain.

A gluten-water dough has a smaller elastic modulus than does a flour-water dough. This appears to show that starch is a not an inert (noninteracting) ingredient in flour-water doughs. Instead starch appears to act as a filler in the gluten polymer. Filled polymers generally are known to have a larger modulus than their unfilled counterparts. This situation is similar to using carbon black as a filler in the rubber for tires. This increases the modulus of the now "filled" rubber, resulting in longer tire wear.

Batters

A cake batter is a very complex mixture of interacting ingredients. While basically an aqueous system, it has a number of dispersed phases. For example, the fat, air, and starch granules are all dispersed phases in the aqueous system. If the viscosity of the batter is too low, these phases separate readily, leading to cakes with a tough, rubbery layer (the gelatinized starch) at the bottom of the pan and a light open-cell structure (the air) at the top. On the other hand, while proper viscosity stops the separation (more accurately, slows it to where it appears stopped in the time needed to make a cake), it does not, in itself, assure a good quality cake.

The first step in producing a batter is mixing. During mixing, the soluble components of both the flour and the added ingredients are dissolved. The insoluble components hydrate. Another important factor at this point is the incorporation of air into the batter. After the batter is mixed, subsequent leavening and baking can form no new air cells. On the other hand, air cells can be lost from the batter either by rising to the surface and being lost or by two cells coalescing into one.

Because of the many dispersed phases in a batter, it is difficult to obtain a true measure of the batter's viscosity. Most procedures to measure batter viscosity call for the batter to undergo severe shearing by being stirred or poured. Because the batter is a non-Newtonian material, its viscosity changes as a consequence of the measurement being made. Thus, we are measuring how the batter changes as it is sheared and not the true viscosity of the batter. This is, usually, not what is desired.

OSCILLATORY PROBE RHEOMETER

With an oscillatory probe rheometer, the probe is immersed in the sample to a prescribed depth (Fig. 8). The probe oscillates at a high frequency but with a very small amplitude. The sample provides a surface load on the probe, and, thus, more power is required to maintain its amplitude and frequency of oscillation. The power required is related to the complex viscosity of the sample. In some instruments, the oscillation frequency is also allowed to vary. This allows the calculation of the rheological parameters G', G'', and tan δ with the appropriate equations.

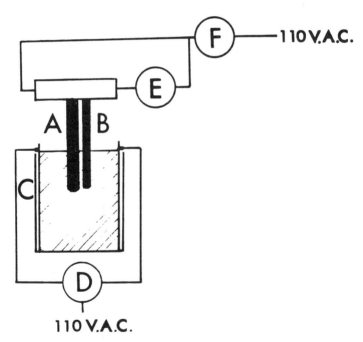

Fig. 8. Schematic diagram of the oscillatory rheometer and electrical resistance oven combination. Viscometer probe (A), temperature probe (B), electrical resistance oven (C), auto transformer (D), isolation transformer (E), and a two-pen recorder (F).

The above procedures provide a constant viscosity reading that is used to study cake batters. As shown in Fig. 9, the apparent viscosity of the batter does not change as a function of the measurement time. One can argue that the batter is, in fact, being sheared at the probe's surface and that, at the high frequency, viscosity has become constant. However, if this were true, one would expect the batter's apparent viscosity to show a decrease when the probe is first turned on instead of the increase that is seen. Another possibility is that the amplitude of the probe is small enough that it does not shear the batter appreciably. Regardless of which, if either, of the above theories is correct, the observed fact is that the apparent viscosity does not change as a function of measurement time. It is this fact that allows us to measure the effects of various formula ingredients and heating on the batter viscosity.

EFFECT OF BATTER TEMPERATURE

The first problem in studying the effect of temperature is to obtain a constant temperature throughout the batter. This is difficult to accomplish as, generally, large temperature gradients are produced in the batter as it is heated. One solution to this problem is to heat the batter by electrical resistance heating (*ohmic heating*).

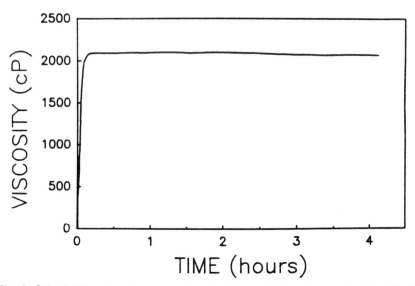

Fig. 9. Cake batter viscosity measured using the oscillatory rheometer at ambient temperature over an extended testing period (4 hr).

If we couple resistance heating with the oscillatory probe rheometer, we can measure the change in viscosity as a function of temperature (Fig. 10). As is well known empirically and shown in the figure, as its temperature is increased, the viscosity of the batter decreases. This continues until the starch in the batter begins to gelatinize, at which point the viscosity increases rapidly.

Conclusions

Rheological measurements have become increasing popular in recent years. This is probably more the result of new and easier-to-use rheometers becoming available than of the increased need for rheological measurements. While rheological measurements are quite useful and often provide us with information that gives a better insight into the

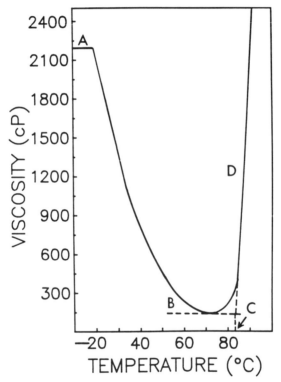

Fig. 10. Viscosity-temperature profile of the AACC cake formula during heating. Viscosity at ambient temperature (A), minimum viscosity of heated batters (B), onset temperature of rapid viscosity increase (C), and rapid viscosity increase (D).

system, they are not an end-all. We cannot determine from our rheological measurements whether our dough will make a good loaf of bread.

REVIEW QUESTIONS

1. What is the study of rheology?
2. Explain the difference between Newtonian and non-Newtonian behavior.
3. What is a modulus?
4. Why is it difficult or impossible to obtain fundamental rheological data from a farinograph or a mixograph?
5. What is the phase angle and what does it tell us?
6. Given the phase angle, how can you calculate G'?
7. Given the phase angle, how can you calculate G''?
8. What does the tangent of the phase angle give and what does it mean.
9. In lay terms, what are G' and G''?
10. Draw an extensigraph curve and explain what the curve tells about the dough under test.
11. Explain why there is a fundamental difference between the information obtained from an alveograph and that from an extensigraph.
12. What is one of the major problems in measuring the rheological properties of fermenting doughs?
13. Explain the lubricated uniaxial compression test and what its major advantage is.
14. What do we mean when we say that dough is viscoelastic?
15. Contrast the rheological properties of wheat gluten and natural rubber.
16. Contrast the rheological properties of wheat gluten and vulcanized rubber.
17. Explain why a gluten-water dough has a smaller elastic modulus than does a flour-water dough.

18. What does it mean when we say we have a filled polymer?
19. Is dough a filled polymer, and, if so, what is the filler?
20. Why is it difficult to measure the viscosity of a cake batter?
21. Explain how a oscillatory probe rheometer works.

SUGGESTED READING

BAGLEY, E. B. 1992. Mechanistic basis of rheological behavior of foods. Pages 573-594 in: Physical Chemistry of Foods. H. G. Schwartzberg and R. W. Hartel, eds. Marcel Dekker, New York.

D'APPOLONIA, B. L., and KUNERTH, W. H., eds. 1984. The Farinograph Handbook. Am. Assoc. Cereal Chem., St. Paul, MN.

FARIDI, H., ed. 1985. Rheology of Wheat Products. Am. Assoc. Cereal Chem., St. Paul, MN.

FARIDI, H., and FAUBION, J. M., eds. 1986. Fundamentals of Dough Rheology. Am. Assoc. Cereal Chem., St. Paul, MN.

FARIDI, H., RASPER, V. F., and LAUNAY, B., eds. 1987. The Alveograph Handbook. Am. Assoc. Cereal Chem., St. Paul, MN.

GRAESSLEY, W. W. 1984. Viscoelasticity and flow in polymer melts and concentrated solutions. Pages 97-154 in: Physical Properties of Polymers. J. E. Mark, A. Eisenberg, W. W. Graessley, L. Mandelkern, and J. L. Koenig, eds. American Chemical Society, Washington, DC.

RAO, V. N. M. 1984. Dynamic force deformation properties of foods. Food Technol. 38(3):103-109.

RASPER, V. F., and PRESTON, K. R., eds. 1991. The Extensigraph Handbook. Am. Assoc. Cereal Chem., St. Paul, MN.

SHOEMAKER, C. F., LEWIS, J. I., and TAMURA, M. S. 1987. Instrumentation for rheological measurements of foods. Food Technol. 41(3):80-84.

SPERLING, L. H. 1986. Introduction to Physical Polymer Science. Wiley-Interscience, New York.

Yeast-Leavened Products

CHAPTER 12

Wheat is grown on more land than is any other food crop. The reasons for that fact are probably twofold. First, the wheat plant is quite hardy and can grow under a variety of environmental and soil conditions. Second, many people like the products that are made from wheat. They like the taste, and particularly, the texture of wheat-based products.

Wheat is unique among the cereals in that its flour possesses the ability to form a dough when mixed with water. In addition, wheat doughs have the ability to retain the gas produced during fermentation or by chemical leavening and thus give a leavened product. These two characteristics of wheat flour doughs are responsible for our preference for wheat products.

This chapter deals with yeast-leavened products made from doughs. By its very nature, the discussion is restricted primarily to wheat flour systems. Rye flour is also used to make yeast-leavened products in some parts of the world, but in the United States, yeast-leavened products made from 100% rye are rare. Rye tends to be used as a flavoring agent rather than a basic dough ingredient. It is unusual to find yeast-leavened prducts made from other cereal flours anywhere in the world.

The most popular yeast-leavened product by far is bread. The amount of bread consumed in the world is truly staggering. Also staggering is the wide array of sizes, shapes, textures, and tastes that bread comes in (Fig. 1). For example, breads vary in size from small bread sticks to loaves weighing several kilos. Crust color and texture can vary from the thick, black crust of pumpernickel to the thin, white crust of Chinese steam bread. The reasons for the large variation are complex and difficult to sort out. Many of them have to do with tradition; one likes the bread that one was raised on. Also important are the other foods in the diet, how much of the diet is bread, and many other factors. Two factors of great importance in the United States, convenience and

230 / CHAPTER TWELVE

economics, could be easily overlooked. Most Americans who return from their first trip to Europe rave about the bread they enjoyed there. When they are told that such bread is available in most U.S. cities at small retail bake shops, they may drop by occasionally for a short time. However, they soon revert to the supermarket shelves. Supermarket bread is more convenient and less expensive than bread from retail bake shops. Therein lies the difference between U.S. bread and the bread that is produced in most of the rest of the world. In most of the world, bread is consumed within a few hours and most certainly within the first day after it is produced. Much of the bread that is produced is truly inedible the day after baking. In the United States, by contrast, bread may not reach the supermarket shelves within the first 24 hr after baking. In addition, the bread must remain soft and edible for six to seven days after baking. It is no wonder that the breads produced in the United States are quite different. Another factor that may not be appreciated is the truly outstanding quality of the bread wheat grown in the United States and Canada. The only other

Fig. 1. Middle Eastern types of bread: Tunisian bread (upper left), Indian *naan* (upper right), Iranian *barbari* (lower left), and *lahvosh* (lower right).

places where wheats of such bread-making quality are produced are in parts of Australia, Argentina, the U.S.S.R., Hungary, the Middle East, and the Punjab area of India. With the wheats produced in most of the world, it would be impossible to produce the high-volume, soft loaf with which the United States is familiar.

Bread-Making Systems

Bread is made by many different procedures. The particular procedure used depends upon many factors, including tradition, the amount (cost) and type of energy available, the type and consistency of the flour available, the type of bread desired, and the time between baking and eating.

FORMULA

The minimum formula for bread is flour, yeast, salt, and water. If any one of these ingredients is missing, the product is not bread. Other ingredients that are often found in the formula are fat, sugar, milk or milk solids, oxidants, various enzyme preparations (including malted grain), surfactants, and additives to protect against molds. Each of the components in the formula performs a function in producing the finished loaf.

The flour, of course, is the major component and is responsible for the structure of the bread. It allows the formation of a viscoelastic dough that retains gas. The role and quality attributes of bread flour are discussed later in this chapter. Yeast is one of the fundamental ingredients; its major role is to convert fermentable carbohydrates into carbon dioxide and ethanol. The gases that result from that conversion provide the lift that produces a light, or *leavened*, loaf of bread. In addition to its gas production, the yeast has a very marked effect on the rheological properties of dough. This effect is also discussed in more detail later in this chapter. Salt is generally used at levels of about 1–2% based on the flour weight. It appears to have two major functions. First is taste; bread made with no salt is quite tasteless. The second is to affect the dough's rheological properties. Salt makes dough stronger, presumably by shielding charges on the dough proteins. The last fundamental ingredient is water, which is a plasticizer and solvent. Without water, we have no dough and therefore no viscous-flow properties, and many of the reactions that take place during fermentation cannot occur because there is no solvent.

In bread that is to be stored for any significant period of time after baking, shortening is an essential ingredient. Bread containing fat in the formula stays soft and more palatable for a longer period of time than does bread prepared without shortening (Fig. 2). In addition to its antistaling properties, shortening has other functions in bread baking. It gives bread with increased volume compared to bread made without shortening. The increase in volume is significant, usually about 10%. The mechanism of the volume increase is still being studied. Fat or shortening also acts as a plasticizer in dough. Thus, if one increases the amount of shortening in the dough, one must decrease the amount of water and vice versa to maintain equal dough consistency. Sugar also is added to the formula, for two reasons. It is a source of fermentable carbohydrate for the yeast, and it provides a sweet taste to the bread.

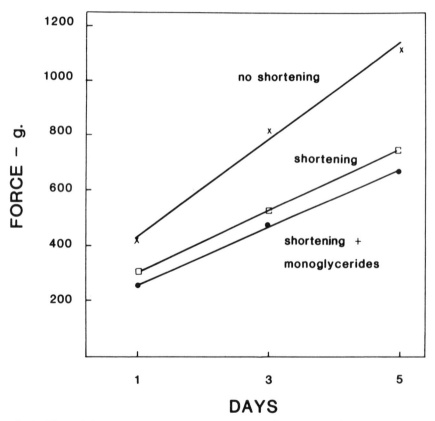

Fig. 2. Effect of shortening and monoglyceride on the firming rate of bread.

With the proper enzymes in the dough, sufficient sugar is produced from the flour to maintain fermentation, and added sugar is not necessary for gas production. However, under most production conditions, added sugar is used for fermentation. Most people will state, when asked, that they do not prefer sweet bread. However, if they are given a series of breads to taste and instructions to rank the breads according to flavor, their rankings generally follow the sugar level. This probably explains why the level of sugar in U.S. bread continues to increase.

Milk or milk solids were a common ingredient in U.S. bread a few years ago. The nutritional advantages of milk are well known. It also had widespread consumer appeal and, because the United States was producing a large excess of milk, it was a relatively cheap ingredient. The supply situation has now changed dramatically, and milk is an expensive ingredient. Therefore, it is not used much, if any, today. More common are milk replacers or whey mixtures.

Oxidants such as ascorbic acid, potassium bromate, azodicarbonamide, and calcium peroxide, at levels of parts per million, improve dough strength and result in bread with better loaf volume and texture. How the oxidants function is discussed later.

In sound wheat flour, the level of α-amylase is very low, and it is common practice in the United States to add malted wheat or barley flour to bread flour. The addition of malt increases loaf volume to a small extent, improves bread texture, and decreases the "keyholing" phenomena. *Keyholing* is the shrinking of the bread sidewalls to give the bread a keyhole shape. The often-stated advantage of having malt produce sugar is true but is only of importance if no, or little, sugar is added in the formula. In recent years, the trend has been to use fungal α-amylase rather than malt flour. The economic advantages are obvious. However, the thermal stabilities of the two types of α-amylase are quite different. This would appear to be important if the desired action of the amylase is on the starch gelatinized during baking. Recently, a trend toward using bacterial α-amylase has developed. It appears to have advantages in keeping the bread from firming during storage.

Bread produced in the United States that must remain soft for a number of days usually contains a surfactant. Most common compounds for this usage are α-monoglycerides; a common level of usage is 0.5% based on the flour weight (Fig. 3). Other surfactants are used as *dough strengtheners*. These are helpful in allowing the dough to withstand the mechanical abuse of the bakery's processing lines. Examples of this type of surfactant are sodium stearoyl lactylate (SSL), ethoxylated monoglycerides (EMG), diacetyl tartaric acid esters of mono- and diglycerides (DATEM), and others (Fig. 3). They also are used at about 0.5% of the flour weight. The additive most commonly used to stop mold growth is calcium propionate.

PROCESSING

The processing of bread can be divided into three basic operations: mixing or dough formation, fermentation, and baking. The simplest bread-making procedure is a *straight-dough system* (Fig. 4). In such a system, all the formula ingredients are mixed into a developed dough that is then allowed to ferment. During its fermentation, the dough is usually punched one or more times. After fermentation, it is divided into loaf-sized pieces, rounded, molded into the loaf shape, and placed into the baking pan. The dough is then given an additional fermentation

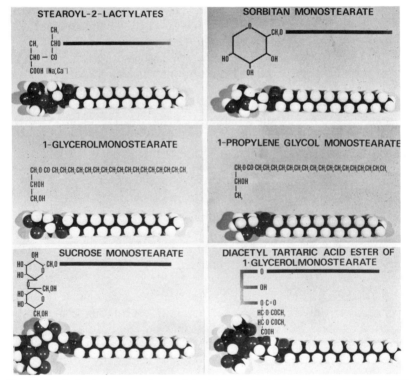

Fig. 3. Surfactants commonly used in baking. Propylene glycol monostearate is commonly used in cake mixes, the others in breadmaking.

(a *proof*) to increase its size. After reaching the desired size, it is placed in the oven and baked. In the straight-dough system, the fermentation time may vary quite widely, from essentially no time to as long as 3 hr.

In general, straight-dough bread is chewier than bread made by other techniques; it has a coarser cell structure; and it is generally considered to have less flavor. The quality of the procedure is quite sensitive to the timing between individual process steps. With larger batches, its time sensitivity can be a problem: if the first of the batch receives optimum fermentation, the end of the batch becomes quite overdeveloped (fermented too long).

The most popular baking process in the United States is the *sponge-and-dough* procedure (Fig. 5). In this procedure, part of the flour (approximately two thirds), part of the water, and the yeast are mixed just enough to form a loose dough (*sponge*). The sponge is allowed to ferment for up to 5 hr. Then it is combined with the rest of the formula ingredients and mixed into a developed dough. After being mixed, the dough is given an intermediate proof (referred to a "floor time") of 20–30 min so that it can relax, and then it is divided, molded, and proofed as is done in the straight-dough procedure.

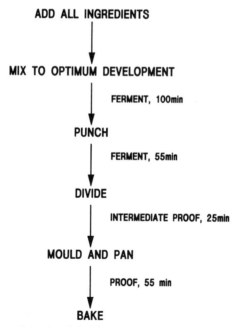

Fig. 4. Outline of a straight-dough baking process.

The sponge-and-dough procedure gives a soft bread with a fine cell structure. It is generally considered to have well-developed flavor and is the basis for comparison for U.S. breads. One of the great advantages of the sponge-and-dough procedure is its tolerance to time and other conditions.

The numerous other bread-making systems that exist can be viewed as either modifications of one of the above two procedures or as new procedures, depending upon one's point of view. Among the other procedures are the liquid sponge, or *preferment*, systems, where the fermentation is performed on a liquid in a tank instead of on a sponge. In this case, part or all of the flour is held out of the fermentation. Such procedures appear to be based upon the assumption that the metabolic products resulting from fermentation are what is beneficial to the dough properties. This is apparently not true (see section on fermentation) and may be responsible for the limited success of these procedures. The *continuous bread-making procedure* that became popular (and was used for >40% of production) in the United States a few years ago was, in part, such a procedure (Fig. 6). It used a preferment, after which the dough was mixed into a developed dough and extruded into the pan, proofed, and baked. The procedure was economical; fewer

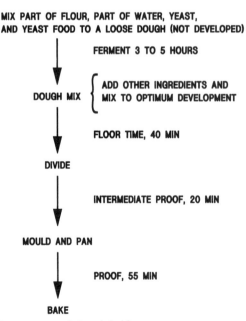

Fig. 5. Outline of a sponge-and-dough baking process.

personnel and less time were required to produce the same amount of bread. However, the bread produced was different from sponge-and-dough bread and not well accepted by consumers. The procedure is essentially no longer used.

Short-time baking systems have became popular in England and Australia. In the United Kingdom, the Chorleywood procedure is used for about 80% of total bread production. This procedure mixes the dough under a partial vacuum, after which it is essentially a no-time (no fermentation time) straight-dough system. It is economical and produces a bread that is desirable in the United Kingdom. In Australia, a short-time procedure is also used; however, it is different from the Chorleywood. It is also a no-time procedure but uses more chemicals (generally oxidants) for development.

Dough Formation

Most food systems containing wheat flour are started by mixing flour, water, and various other ingredients to form a dough. A dough is more

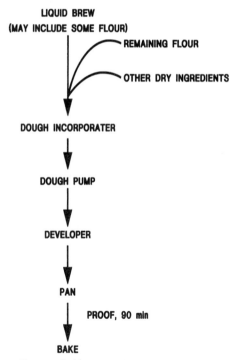

Fig. 6. Outline of a continuous-baking process.

than just a flour-water system. When wheat flour and water are mixed in various proportions, they form everything from a slurry when water is in large excess, to a dry but slightly cohesive powder when flour is in large excess. At intermediate levels, they are more likely to produce a sticky mess. When such a system is continuously stirred or mixed, remarkable changes occur. The system appears to become less wet and sticky, and a dough that is cohesive and partially elastic is formed. Such a dough is resistant to extension, as is shown by a mixograph curve (Fig. 7). As the dough is mixed for longer periods of time, it becomes more resistant to extension (i.e., the height of the curve is greater). The dough then is said to be "developed".

What do we mean by dough development? To start with, flour, particularly hard wheat flour, is made up of discrete particles (Fig. 8). Although we think of the particles as small, they are actually quite large (100–200 μm) compared to starch granules (5–40 μm) or, especially, protein molecules. When water is added to the flour particles, the surfaces of the particles rapidly hydrate, because water is in large excess compared to the surface area of the particles. Because of the large excess of water, the system is quite fluid and not much resistance to extension occurs.

MIXING

Hard wheat flour particles are dense, and water penetrates the particles slowly. The only driving force for the water to move to the

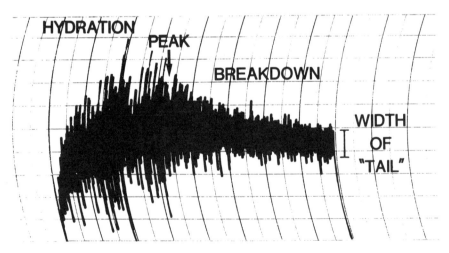

Fig. 7. A mixogram curve labeled to show the various parts of the curve.

center of the particle is diffusion, which is slow. Mixing, however, provides an additional mechanism. As the hydrated particles are rubbed against each other, the mixer bowl, or the mixer blades, the hydrated surface is removed, exposing a new layer of particles to the excess water in the system. As this is repeated many times, the flour particles slowly become completely worn away or hydrated. As more and more of the free water is used to hydrate the protein and starch, resistance of the system to extension is increased progressively. Thus, the height of the mixing curve gradually increases to a peak (Fig. 7).

A dough that has been mixed to a peak can be referred to by a number of terms, for example, *a mixed dough, a dough with minimum mobility,* or *an optimally mixed dough.* All of these imply that an end point has been reached. They also imply that this is the point to which a dough should be mixed for producing a loaf of bread. This raises a number of questions. For example, why is a peak obtained and why is it the optimum? It appears obvious that a peak or plateau occurs because all the flour particles are now hydrated. If they were not all hydrated,

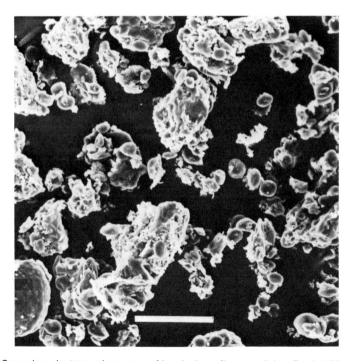

Fig. 8. Scanning electron microgram of hard wheat flour particles. Bar is 100 μm.

then continued mixing would give more resistance to extension and shift the maximum point on the curve to a longer mixing time. The answer to the second question also appears obvious. This is the optimum amount of mixing because all the protein and starch are now hydrated. Protein or starch that are not hydrated cannot interact in the dough in any beneficial way. If we examine the foregoing scenario, we see that dough development is essentially just the result of complete hydration of the flour particles. A scanning electron micrograph of a freeze-dried, optimally mixed dough (Fig. 9) shows no intact flour particles but instead an apparently random mixture of protein fibrils with adhering starch granules. This is a good model for mixed dough.

What, then, do we know about dough development? It is possible that development occurs as the dough is mixed to obtain optimum

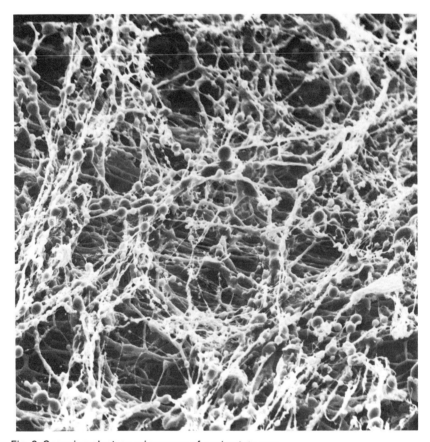

Fig. 9. Scanning electron microgram of a mixed dough.

hydration. One way of examining this is to take advantage of the dough's tendancy to relax. If we mix a dough to optimum, then allow time to pass after mixing, the dough relaxes. When we start to mix it again, the dough does not immediately give a height equal to the peak height that occurred during the initial mixing. Some mixing is necessary to restore the resistance to extension to its original value. It does, however, reach that point with additional mixing. The mixing required could be viewed as the energy necessary to develop the dough.

If the above is true, then development is a reversible process. We can develop a dough, allow it to relax, and then develop it again. This implies that the bonding involved is not covalent but probably hydrogen or hydrophobic bonding or both.

Another important aspect of dough mixing is the incorporation of air. As dough becomes cohesive, it starts to incorporate air and thus decreases in density. At the point of optimum mixing, about half the total amount of air that is possible to incorporate has been incorporated. This air, particularly nitrogen, is important in most baked products because it produces the cells to which the CO_2 diffuses. Yeast cannot produce new gas cells. Because of this fact, if there were no air cells, the grain would be very coarse with only a few large cells. Stated in different terms, the nitrogen gas trapped during mixing provides the nuclei for subsequent gas expansion, or leavening of the dough.

Some doughs (in the Chorleywood method, for example) are mixed under vacuum. At first glance, this would appear to be detrimental to their quality. However, the vacuum is not complete. High vacuum cannot exist because dough contains water. Actually, the partial vacuum is beneficial because the small bubbles created by the mixing expand under the reduced pressure and thus can be subdivided into more bubbles as the mixing proceeds. The more bubbles created, the finer the grain of the baked product will be.

MIXING TIME

An examination of mixing curves, such as those shown in Fig. 10, clearly shows that the mixing peak for various flours occurs at different times. The area under the curve is essentially a power curve, i.e., a measure of the work required to mix a dough. Obviously, to reach optimum development, different flours should not be mixed to the same work input.

The mixing time of various wheat flours is under genetic control and, like any other character, can be selected for by breeders. The protein content of the flour also can affect mixing time. It has been shown that low-protein flours (<12% protein) require longer mixing times simply because they contain less protein. Protein content above 12% does not affect mixing time. Another factor affecting mixing time is the environment under which the wheat was grown. The control of mixing time appears to be associated with the the glutenin protein fraction of wheat flour.

Certain chemical agents, particularly reducing agents such as cysteine, sodium bisulfite, and related compounds, are quite effective in shortening mixing time. These reagents apparently work by breaking disulfide bonds in the glutenin proteins, thus making the proteins smaller. These smaller proteins hydrate more easily and thus lead to shorter mixing times. The pH of the dough also affects mixing time, with lower pH giving a shorter mixing time and higher pH (up to pH 10) giving longer times. This can be explained as an effect upon the charge of the proteins. The effect of low pH can be overcome with salt (presumably by charge shielding).

OVERMIXING

After reaching the optimum development, continued mixing produces a wet, sticky dough with an "overmixed" sheen. This is generally spoken

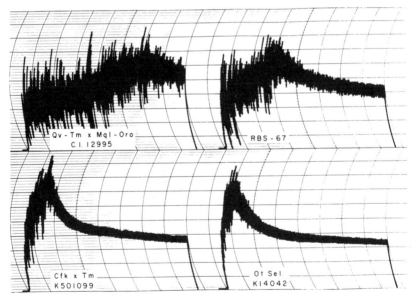

Fig. 10. Mixogram of hard winter wheats showing excellent (CI 12995), good (RBS-67), poor (K501099), and extremely poor (K14042) mixing properties.

of as the dough being *broken down*.

One explanation of overmixing is that shear-thinning occurs. If we continue to mix the long protein molecules, they line up in the direction of flow and thus offer less resistance to mixing. Although this is a good explanation, it apparently is not correct because doughs do not overmix in a nitrogen atmosphere or if the water-soluble fraction is removed. There is no reason to believe that either of these two factors would influence shear-thinning. Thus, we must look for another explanation.

The fact that doughs mixed in a nitrogen atmosphere do not overmix implies that overmixing is an oxidation process. The fact the doughs do not overmix if the water solubles are removed implies that something in the water-soluble fraction is involved in the process. The clue to what was important in the water-soluble fraction came from the finding that fumaric acid and related compounds reduced mixing time and also greatly increased the rate of dough breakdown. Certain α- or β-unsaturated carbonyl compounds, such as fumaric acid, maleic acid, or ferulic acid have effects on overmixing (Fig. 11) similar to those of the sulfhydryl (SH)-blocking reagent N-ethylmaleimide (NEMI). However, NEMI and cysteine interact during dough mixing, whereas fumaric acid and cysteine do not. Thus, the effect of the activated double-bond compounds cannot be explained by the SH-blocking.

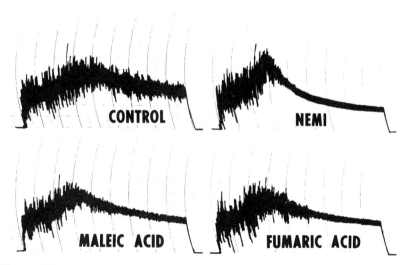

Fig. 11. Mixograms showing the effect of activated double-bond compounds. N-ethylmaleimide (NEMI) at 500 ppm; fumaric and maleic acids at 2,000 ppm.

Surprisingly, the effect of the activated double-bond compounds on dough breakdown during overmixing can be reversed by lipoxygenase (e.g., enzyme-active soy flour) or by free radical scavengers. This fact led to the hypothesis that the activated double-bond compounds have their effect by reacting with free radicals created in the gluten protein during dough mixing.

Interestingly, fast-acting oxidants, such as potassium iodate and azodicarbonamide, induce a rapid dough breakdown that appears very similar to that produced by activated double-bond compounds. One explanation of the effect of fast-acting oxidants is that they oxidize SH groups in the flour to disulfides. Because of the resulting lack of SH groups, no thiol-disulfide interchange reactions occur between protein molecules. These theoretically could reduce the stress on large glutenin molecules as they are sheared (mixed). Thus, the gluten proteins become more stressed, which results in more disulfide bonds being broken, forming more thiyl radicals that could react with the activated double bonds. Therefore, breakdown occurs more rapidly even though the number of activated double bonds remains the same. An alternate and perhaps better explanation is that the strong (fast-acting) oxidants oxidize something in the flour to produce additional activated double-bond compounds. To date, such compounds have not been identified.

Fermentation

Yeast is a living organism that is inactive during storage. The inactivity is caused either by drying, in the case of active dry yeast, or by low temperature, in the case of compressed or crumbled yeast. Yeast, as it is produced commercially, is always contaminated with bacteria, mainly lactobacilli, which are quite important in crackers and sourdough bread but do not appear to be important in regular bread processes.

When yeast is incorporated into a dough, conditions are suitable for it to become active. Yeast is a versatile organism; it can ferment under either aerobic or anaerobic conditions. The production of yeast and the early stages of brewing are aerobic processes, whereas bread fermentation is an anaerobic process. Thus, little growth of yeast occurs during dough fermentation. The oxygen in a dough is rapidly consumed by the yeast and bacteria as fermentation starts. Thereafter, the fermentation is anaerobic unless we add oxygen to the system (i.e., by remixing). The major products of yeast fermentation are carbon dioxide and ethanol. As carbon dioxide is produced, the pH decreases and the aqueous phase becomes saturated with carbon dioxide. The initial lag that is found in a gas "production" curve for bread dough is because the dough's aqueous phase must become saturated with carbon dioxide before the evolution or loss of CO_2 can be measured. Only after the aqueous phase has become saturated is the carbon dioxide available to leaven the system. The leavening process itself is discussed in the next section. At this point, our focus is on the obvious fact that gas is retained and the dough is leavened.

As fermentation proceeds, it is customary to punch or remix the dough, depending upon which baking system is being used. Why is this done and what does it accomplish? The gas cells in the dough become larger and larger as more gas is produced. Punching or remixing subdivides the gas cells to produce many more smaller cells. To be sure, a large amount of carbon dioxide is lost to the atmosphere, but the important aspect of the process is the creation of the new gas cells.

Another important benefit of punching or remixing is the mixing of the dough ingredients. Yeast cells do not have mobility in dough. Therefore, they depend upon the sugar diffusing to them. As fermentation proceeds, the diffusion distances become large, so the concentration of sugar diminishes, along with the rate of fermentation. Punching or remixing brings the yeast cells and fermentables together again. In zero- or short-time baking systems, punching is not practical, as the dough is not given sufficient time to expand. The net result is usually a coarser grain (fewer cells) in the bread. A partial solution to this problem is to mix under partial vacuum, which expands the dough and allows the gas cells to be subdivided without the need for waiting for the dough to expand.

In addition to its gas-producing capabilities, yeast also affects dough rheology. The effects of yeast on dough rheology can best be shown by a simple *spread test*. The logic of the test is shown in Fig. 12. One can consider a dough to have both viscous-flow properties and elastic properties. A dough that has more viscous-flow properties has a large *spread ratio* (width divided by height), whereas a dough that is more elastic has a smaller one.

As can be seen in Fig. 13, a flour-water dough gives a large spread ratio after 3 hr. This indicates that the viscous-flow properties are large in a flour-water dough. When yeast is added to such a dough, the spread ratio is quite different. This shows that yeast influences the rheological properties of a dough. The addition of yeast to the formula causes a dough to go from one with a large viscous-flow component to one that is elastic, as a result of 3 hr of fermentation.

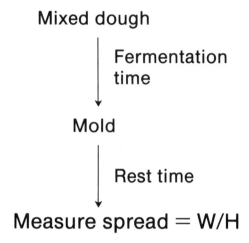

Fig. 12. Experimental scheme for determining the "spread" of a wheat flour dough. W, width; H, height.

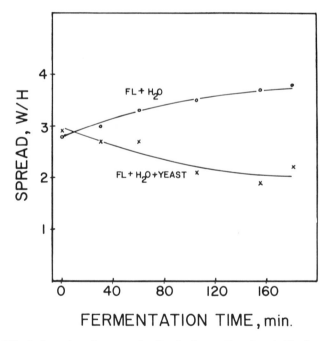

Fig. 13. Effect of yeast on the spread ratio of a fermenting dough. FL, flour.

The trend toward a dough with more elastic properties is the same trend that we find when we add oxidants to a dough. Thus, yeast clearly has an oxidizing effect.

This brings up an obvious question. Do the products of fermentation produce the rheological change or is this a property of the yeast itself? The question can be easily answered by running a preferment containing no flour and centrifuging the system to separate yeast cells from the products of fermentation. When this is done and each is added to separate flour-water doughs, it is clear that the products of fermentation do not change dough rheology. The yeast itself appears to be the entity that changes dough rheology. This simple experiment

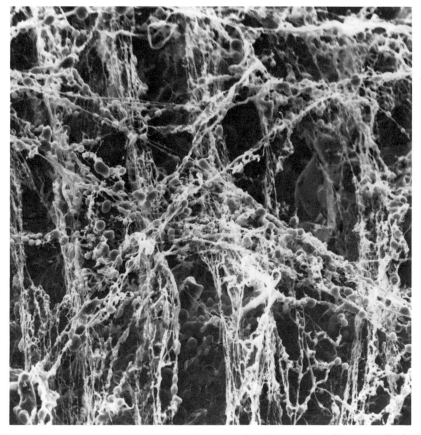

Fig. 14. Scanning electron microgram of dough after fermentation. Note the aligning of the gluten fibrils.

probably explains why preferment systems containing no flour have not been successful. How the yeast cell changes dough rheology is not clear and is an area that needs to be studied in more detail.

The chemical oxidants that are added to the bread formula also affect dough rheology. Certain of the oxidants (potassium iodate and azodicarbonamide, for example), are fast-acting oxidants that have their effect during mixing. Potassium bromate, on the other hand, has essentially no effect during mixing but does affect dough rheology during fermentation. At least part of potassium bromate's time-dependent effect may be because of the change in dough pH during fermentation; potassium bromate reacts faster at lower pH. Ascorbic acid has both a rapid and a time-dependent reaction on dough rheology.

An optimally fermented and oxidized dough has no viscous-flow properties under the force of gravity at the proof stage. Thus, the dough in the pan expands to fill the pan rather than flows to fill the pan. After fermentation, presumably because of the mechanical punching, the gluten fibrils appear to be aligned (Fig. 14).

The change in pH associated with fermentation time is also important to the dough's rheological properties. Dough just out of the mixer usually has a pH of about 6.0. During fermentation, the pH drops to 5.0. A first rapid drop is caused at first by carbon dioxide dissolving in water to produce carbonic acid. A second factor is the slow production of organic acids by the bacteria in the dough. The flour itself and either milk or soy proteins in the formula are good buffers and therefore help to control pH. The lower pH decreases the mixing time of dough. This is, at least in part, the reason for the shorter mixing time in a sponge-and-dough or preferment system than in a straight-dough system. However, the change in pH has little effect on the dough's spread ratio.

GAS PRODUCTION

The process of anaerobic fermentation of carbohydrates by yeast to the end products, ethanol and carbon dioxide, is summarized below:

$$C_6H_{12}O_6 + \text{yeast} \longrightarrow 2C_2H_5OH + 2CO_2$$
$$\text{Glucose} \phantom{+ \text{yeast} \longrightarrow 2} \text{Ethanol} \quad \text{Carbon dioxide}$$

In a system designed to maximize yeast activity, the substrate (sucrose or glucose) is utilized at a rate of 0.77–3.0 g of sugar per hour per gram of yeast solids. In a 3-hr fermentation and 55-min proof, with compressed yeast (29% solids) added at 2% flour weight, that calculates to be 1.75–6.82 g of sugar. Estimates of the sugar consumed, based on the fermentation products and the volume of CO_2 necessary for dough leavening, have been reported as approximately 3.5% sugar (based on the flour weight) utilized during the 4-hr fermentation and proof. This value, however, does not consider loss of CO_2 to the atmosphere. Other estimates have been made that during fermentation of a straight dough, approximately 2.0% sugar (based on flour weight) is consumed from mixing to the end of proof. That value accounts only for the disappearance of sucrose and does not include fermentable carbohydrates naturally present in the flour. Sucrose depletion by yeast in a synthetic fermentation broth during 3 hr is about 4%. In a straight dough, the yeast is in a stressed environment (insufficient water) and would not be expected to maximize substrate utilization.

Carbon dioxide is retained in the bread dough in two phases: 1) as a gas contained within the gas cells and 2) dissolved in the aqueous phase. Carbon dioxide solubility in water is inversely related to the temperature and affected by pH. At the pH of bread doughs, most of the carbon dioxide is present as CO_2 and little of it is as H_2CO_3, HCO_3^-, or CO_3^{-2}.

The amount of carbon dioxide present in a fully proofed dough is only about 45% of the total produced by fermentation. The balance is lost during fermentation, punching, molding, and proofing. The expansion of CO_2 in the air cells and the CO_2 that comes out of the aqueous phase are insufficient to completely account for the increased volume of dough in the oven. Vaporization of the water-ethanol azeotrope (boiling point 78°C) during heating in the oven contributes greatly to the overall expansion of the dough.

GAS RETENTION

Why does a wheat flour dough retain gas? This appears to be an important question that must be answered if we are to understand the dough system. Research showed years ago that yeast could create no new bubbles in a dough system. Bubble mechanics shows that the pressure (P) in a bubble is related to the radius of the bubble (r) and the interfacial tension (γ) by the following relationship:

$$P = \frac{2\gamma}{r}$$

Thus, in a system where the interfacial tension does not change, if r approaches zero, then the pressure required to start a new bubble is infinite. Where, then, do the bubbles come from? We must incorporate air during mixing to provide preexisting bubbles. Air, actually nitrogen, appears to be the gas of choice to do this because it is not very soluble in water. Both carbon dioxide and oxygen dissolve in water and thus do not provide the needed bubbles. As shown in Fig. 15, air is incorporated as the dough is mixed. At optimum mixing, only about half the total possible air is incorporated. To make the grain of the product finer, a number of things can be done. We can ferment the dough for a time and subdivide the now-larger bubbles by punching or remixing the dough; this gives more bubbles and thus a finer grain. We can mix under a partial vacuum; the reduced pressure makes the bubbles larger so they can be subdivided into more bubbles. We also can change the size of the bubbles formed during mixing by changing the interfacial tension of the system. Surfactants change the interfacial tension and thus affect the grain of bread.

To sum up this point, in a mixed bread dough, an insoluble but highly hydrated gluten protein system constitutes the continuous phase, with starch and air bubbles as the discontinuous phases (Fig. 16). Also dispersed throughout the aqueous system are yeast cells, which ferment sugar and produce, among other things, carbon dioxide. The CO_2 is produced in the aqueous phase and saturates the water. Once the water is saturated, newly produced CO_2 must find a place to go. As discussed

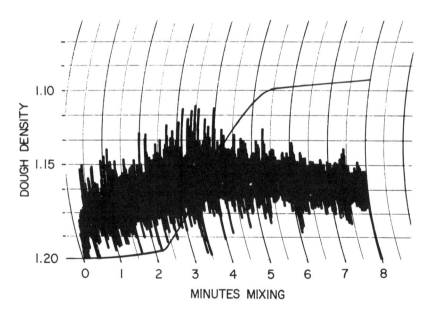

Fig. 15. Mixogram showing the decrease in dough density as the dough is mixed.

above, it cannot form new bubbles; thus, the preexisting air bubbles appear to be a logical choice. The CO_2 enters the bubble and increases the pressure. Dough has viscous-flow properties and therefore allows the bubble to expand to equalize the pressure. The total volume of the dough mass is increased or, in other words, the dough is leavened.

Cereal chemistry dogma holds that the carbon dioxide is retained by the gluten proteins, which form a sheet or membrane. The model that is often used is a rubber balloon. This concept must be questioned. If there is a barrier to gas coming out of the bubble, then how does gas breach the barrier to get into the bubble? Consideration of that question quickly brings us to the conclusion that gluten does not form a membrane or any kind of barrier. In fact, no barrier is needed, as the CO_2 in the bubble cannot diffuse out because the aqueous phase surrounding the bubble is saturated with CO_2 and the yeast is producing more to see that it stays saturated. Gas retention, then, is not a mystery but only an application of the laws of diffusion.

ROLE OF PENTOSANS

Wheat flour contains gums, the water-soluble and water-insoluble pentosans. The water-soluble fraction of wheat flour has been shown to be important in producing an optimum loaf volume. The mechanism of its action is not clear but may be just an increase in the viscosity of the aqueous phase.

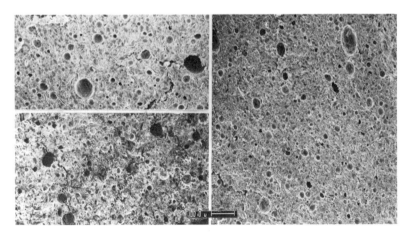

Fig. 16. Scanning electron microgram of cyro-fractured freeze-dried, flour-water doughs. Upper left, control dough; lower left, dough containing 3% shortening; right, dough containing 0.5% sodium stearoyl lactylate.

Rye flour is second only to wheat flour in its ability to retain gas and produce a light loaf of bread. Rye flour contains a much greater amount of pentosans than does wheat flour. Recent studies have shown that the pentosans are the most important fraction in rye flour baking quality.

ROLE OF LIPIDS

If flour is defatted in such a way as not to damage the gluten, the flour still produces a reasonable loaf of bread. However, if the lipids are added back to the flour, the bread volume is larger. Research has shown that the polar lipids, and particularly the glycolipids, are most important in this phenomenon.

ROLE OF GLUTEN

Bread-making data have shown that bread can be made without the water-soluble pentosan faction. The loaf volume is higher when the water solubles are included, but a reasonable loaf is obtained without them. The same is true of the lipids; if the lipids are included, the loaf volume is higher. Thus, although both the water solubles and lipids are clearly important, we must conclude that the gluten proteins are by far the most important.

Much remains to be learned about how a wheat flour dough retains gas. It appears to be time to move beyond the gas-retaining membrane theory. Clearly, if the rate of carbon dioxide diffusion in dough can be slowed, the dough will retain more gas. The structure of the gluten protein results in a slow rate of diffusion. Understanding the structure of the gluten proteins and how they interact with the lipids and pentosans remains a challenging problem.

The gluten proteins govern the bread-making quality of the various wheat flours. Flours milled from different wheat cultivars vary widely in their loaf volume potential, as illustrated in Fig. 17. At a constant protein content, a large variation in loaf volume can still exist. This is what is called *protein quality*. For some reason, the protein from one cultivar is much better at retaining gas than is the protein from another.

The total amount of gluten protein in a flour is also important, as clearly shown in Fig. 17. This is the *protein quantity*. The concept has been readily accepted, but little work has been reported on this mechanism by which greater protein content gives greater loaf volume.

Recent work has suggested that the gluten protein can be extended only to a certain level before reaching its elastic limit. As the loaf expands during baking, that elastic limit and the amount of protein in the dough together determine the size of the loaf.

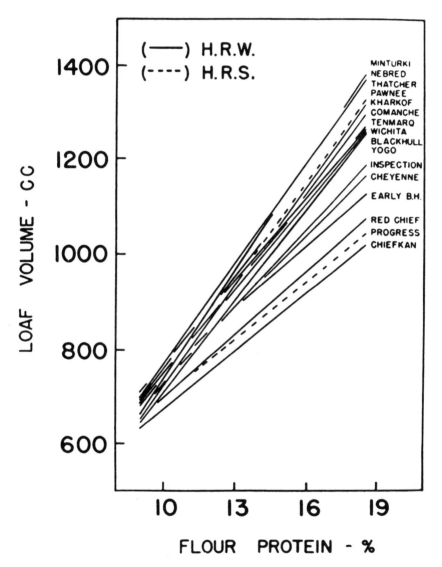

Fig. 17. Effect of flour protein quantity and quality (slope of the lines) on loaf volume. HRW, Hard red winter; HRS, hard red spring; BH, Blackhull.

GLUTEN-FREE BREAD

A number of reports have appeared in the literature purporting to show that bread can be produced in the absence of gluten. One can argue that such products are not bread. Clearly, they have a texture and grain quite different from what we expect from wheat flour. Thus, it is tempting to dismiss such reports, because doughs that retain gas but contain no gluten are troublesome. However, as is often the case, we learn the most from the troublesome data. Apparently, the common thread in these systems is that they slow the diffusion of gas.

Two different phenomena appear to be responsible for the ability of gluten-free systems to retain gas. Some reported systems have contained locust bean gum or xanthan gum, whereas others have used surfactants. The reported data clearly show that those reagents do indeed result in the retention of gas by a gluten-free starch system.

Molding, Proofing, and Baking

After bulk fermentation, the dough is divided into individual loaf-sized pieces and given a floor time. The purpose of this time is to allow the dough to relax. The dough is then ready for *molding*. The molding operation is essentially sheeting followed by curling, rolling, and application of pressure (Fig. 18).

As dough is sheeted (passed between rolls to be flattened) during the various processes, it must be sheeted in different directions. Continued machining in one direction would align the protein fibrils and result in a dough that was strong in one direction but weak in the direction at a 90°-angle to the sheeting. After being molded, the loaf is panned, with the final lap down in the bottom of the pan (Fig. 19).

The dough is then ready for proofing. This is usually accomplished at 30–35°C and at 85% rh. Because the dough now has only limited viscous-flow properties, it fills the pan by expansion. Proofing usually takes about 55–65 min; the dough increases greatly in volume.

After proofing, the dough is ready for baking. Things start to happen rapidly as the relatively cool dough is placed into the hot oven. The surface of the dough that is exposed to the oven atmosphere skins over and forms a crust almost immediately. The skin, or crust, forms because the surface of the dough is rapidly dried. The moisture on the surface of the dough vaporizes very fast, as it is in contact with dry, high-temperature air. The surface of the dough cools as the water vaporizes. Thus, much of the starch in the crust retains its birefringence

YEAST-LEAVENED PRODUCTS / 255

Fig. 18. Molding bread dough from 10 g of flour: initiating (top), curling step (center), and completing (bottom).

in the finished loaf. If a thicker, heavier crust is desired, steam is generally added to the oven. The steam in the oven slows the rate of water vaporization, and the surface cooks to a greater extent, producing a thicker crust. The formation of a crust or skin has nothing to do with the loaf's retention of gas, which is controlled by the properties of the interior of the dough and not by the surface.

Most of the heat that is absorbed by the dough comes through the baking pan. Therefore, the rate at which the dough temperature increases depends upon the heat transfer from the air and the baking surface to the baking pan. Dough does not conduct heat as well as the metal pan; therefore, a well-defined temperature gradient forms from the outside to the center of the loaf during baking.

During the first few minutes in the oven, the dough expands in size rapidly. This is called the *oven-spring*. Several factors are responsible for the oven-spring. Gases heat and increase in volume; carbon dioxide becomes less soluble as the temperature is raised; yeast becomes quite active as the temperature is raised (as long as it doesn't get too high); and other materials (for example, ethanol-water mixtures) are vaporized.

Generally, the oven-spring stage lasts less than 8 min; the remaining baking time is to ensure that the center of the loaf approaches 100°C.

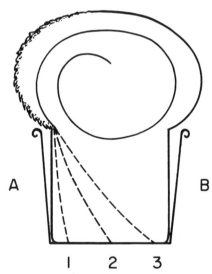

Fig. 19. Cut loaf of bread in pan illustrating configuration of molded dough after expansion and baking. Spiral line represents dough-to-dough interface. Dotted lines represent interface with dough crease at three different positions.

Remember that bread is a moist product; thus, the temperature cannot exceed 100°C unless the product becomes dry. The only part of the loaf that becomes dry is the crust, and this occurs late in the baking cycle. This is why only the crust goes through browning. The other thing that is accomplished with longer baking is the reduction of the moisture content to the desired level.

The grain of a loaf of bread is not a random collection of cells but instead an organized system that can tell us much about the loaf. In an aqueous environment with no added stress, gas cells will always be round. They have the minimum surface area (and, therefore, the least free energy) when they are round. Such cells are clearly seen in cakes or in bread made with the continuous bread-making systems (Fig. 20); therefore, we can infer that there is no external force on the cells in those systems. In conventionally made bread, many of the cells are elongated, indicating that considerable force was put on the cells (Fig. 20). The degree of elongation is, then, a measure of the strength of the dough and a direct measure of the chewiness of the bread.

During molding, the dough is sheeted and then rolled into a cylinder. The cells are elongated around the cylinder. If the dough is strong, the elongation persists through the proof stage. During baking, the crust on the exposed surface of the dough breaks at one or both sides where the crust meets the pan. With a good strong dough, the break occurs on the side of the last lap. As the break occurs, the loaf expands in that direction, giving a break and shred (i.e., an area where new

Fig. 20. Crumb grain of bread produced by the continuous-baking procedure (left) and a sponge-and-dough procedure (right).

dough is exposed to oven temperature as the loaf expands) on the exterior of the loaf and a nice slip plane (i.e., an area where the cells slide past each other as the dough expands) of elongated cells in the interior. If the loaf breaks on the other side, the grain is much less organized and more difficult, if not impossible, to judge. In a fully oxidized and developed (matured) loaf, one would expect a large number of the cells to be elongated, with a distinct slip plane. If the dough is underoxidized ("green"), many small round cells are evident and the slip plane is not distinct. Overoxidized dough gives loaves with larger round cells with thick cell walls and a very prominent slip plane, often containing large open cells. Thus, the grain of the loaf gives a reading on the oxidation requirement of the flour.

Occasionally, reference is found to changing the rate of baking. This is extremely difficult to do. The rate of baking is actually determined by the rate of heat penetration into the dough, and this cannot be changed. Increasing the oven temperature provides a larger gradient of temperature (ΔT), but that only speeds the rate slightly. The higher oven temperature results in larger temperature and moisture gradients in the loaf but will not greatly change the time required to raise the center of the loaf to the desired temperature.

The later stages of baking are when all the browning occurs. Watching a loaf bake in an oven with a glass door is an interesting experience that is highly recommended if one is interested in the process. Browning occurs very late in the system because it takes place much faster in a dehydrated system. The browning on the bread surface results from a Malliard-type reaction. Sucrose is a nonreducing sugar, but yeast contains invertase, which rapidly produces glucose and fructose, both of which are reducing and brown quite readily. Caramelization is occasionally mentioned as a mechanism of browning in bread; however, bread produced by chemical leavening (where there is nothing to convert sucrose to reducing sugar) does not brown to any extent. Thus, the role of caramelization in yeast-leavened products appears to be minor.

Transformation of Doughs into Baked Products

We can all easily tell the difference between dough and bread. However, it is not so easy to understand how dough is converted or transformed to bread. Dough consists of a continuous gluten phase with starch granules and air cells discontinuous in that phase. As discussed earlier, gas is retained by wheat flour doughs, as it can only diffuse slowly through the complex gluten network. When we put dough in an oven, it expands rapidly. This is the well-known phenomenon of "oven-spring." If we assume that we change nothing in the dough by baking it, then the dough should shrink back to approximately its original size after it is removed from the oven. This is what would occur if a rubber balloon were heated and cooled. However, this is not what happens when we remove a loaf of bread from the oven. Our assumption that nothing irreversible has happened to the dough that we placed in the oven is clearly not correct.

The internal structures of doughs that have been in the oven for various periods of time (Fig. 21) show sharp demarcations between the dough and bread regions of each loaf. This line of demarcation is consistent with the gradient of temperature known to develop during baking. Thus, it appears reasonable to assume that temperature is responsible for the transformation of dough to bread.

If bread dough is baked in an electrical resistance oven, the temperature gradients are, essentially, removed and the dough heats uniformly. If we bake dough by this method and, at the same time, measure the carbon dioxide lost from the loaf and the dough temperature, we can determine at what temperature the dough starts to lose carbon dioxide. Figure 22 shows little loss of carbon dioxide until the temperature reaches about 72°C. At that temperature, large amounts of carbon dioxide are released from the dough/bread. Thus,

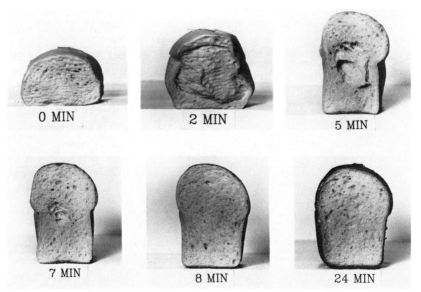

Fig. 21. Cross sections of dough and bread before baking and after 2, 5, 7, 8, and 24 min of baking.

one change brought about by temperature is the loss of the dough's ability to retain carbon dioxide.

The simple experiment of attempting to blow through dough shows that, in dough, the gluten is continuous and the gas is discontinuous (in gas cells). Thus, for example, the dough can be used to blow a bubble (i.e., with the alveograph). On the other hand, a slice of bread offers no resistance to the passage of air. The air freely passes through, showing that bread is not only gluten-continuous but *also* gas-continuous. Therefore, when dough is transformed from dough to bread, it becomes gas-continuous and is no longer capable of retaining gas.

Now that we have an understanding of *what* happens, let's try to understand *how* it happens. When we think of the changes that are caused by heating a dough system, we first think of starch gelatinization. In a bread dough, the starch gelatinizes at about 65°C (Fig. 23). However, note that at the water level present in dough, gelatinization occurs over a wide temperature range. A thermocouple placed in dough as it is baked in the electrical resistance oven shows that the visual transformation from dough to bread occurs at 65°C. However, the physical properties of the dough continue to change as the dough is heated to higher temperatures. Specifically, the bread becomes stronger and more elastic. At 95°C, the bread has lost all of its extensibility and is now elastic. Thus, the transformation of

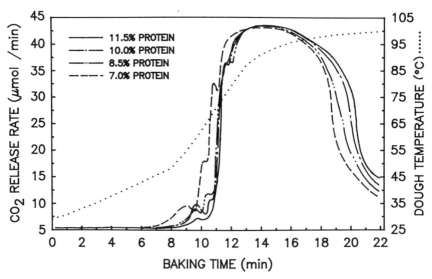

Fig. 22. Effect of protein content on CO_2 release during baking. Dotted line is temperature.

dough to bread is a continuous process. It starts at 65°C; the dough/bread becomes gas-continuous at 72°C; but it continues to become more elastic up to 95°C.

Rheological measurements of dough heated to various temperatures combined with microscopic examination of bread have suggested that there is an interaction of some type between gelatinized starch and gluten. This interaction causes the dough to be more elastic and less extensible. As dough is heated, the reduction in extensibility leads to a buildup of pressure within the gas cell. At some point, the pressure becomes large enough to rupture the gas cell wall and produce the gas-continuous sponge structure that we find in bread. This explanation is supported by the scanning electron micrographs in Fig. 24, in which dough/bread heated to below 70°C shows no tears in the cell walls, whereas dough/bread heated to above 88°C does show tears in its cell walls.

The increase in strength of the bread structure, by itself, is not enough to keep the bread from collapsing or keyholing when it is removed from the oven. If a pressure differential developed when the gas cooled, it would cause the loaf to collapse. However, the fact that the bread is gas-continuous does not allow the pressure differential to develop. Thus, bread does not collapse when it is removed from the hot oven and allowed to cool.

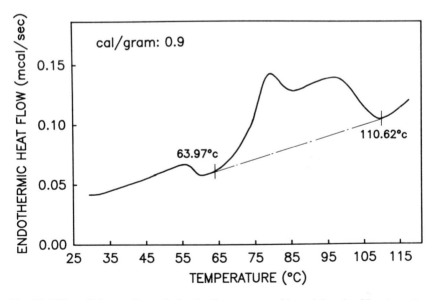

Fig. 23. Differential scanning calorimeter thermogram of bread dough without yeast.

Retrogradation and Staling

"All bread fresh from the oven is good bread." There is a lot of truth in that saying. Bread loses its desirability progressively with the time it is out of the oven. Those undesirable changes that occur with time are collectively called "staling." They include toughening of the crust, firming of the crumb, a loss of flavor, an increase in the opaqueness of the crumb, and a decrease in soluble starch.

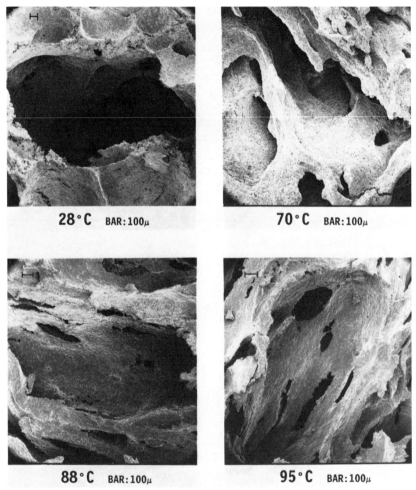

Fig. 24. Scanning electron micrographs of the cross sections of dough and bread baked to 70, 88, and 95°C.

Retrogradation is the recrystallization of the starch. It appears to be mainly the amylopectin fraction that recrystallizes. The increase in opaqueness of the crumb is presumably caused by the growth of the crystallites, which changes the refractive index.

The changes that occur in the crust are clearly different from those that occur in the crumb. The toughening of the crust appears to be mainly associated with migration of water from the crumb to the crust. When bread is freshly baked, the crust is dry, containing 2–5% moisture. Under these conditions, it is friable and desirable. As water diffuses from the crumb, the crust loses its friability and becomes tough, going from a glassy state to one that is leathery or rubbery.

The changes that occur in the crumb appear to be much more complex. It was shown, almost 150 years ago, that the firming of bread crumb is not a drying phenomenon. Firming occurs even though no moisture is lost. Occurring over the same general time span as the firming is a recrystallization of the starch. This is referred to as "retrogradation." Over the last 20 years or so, there has been a general consensus that firming and retrogradation are the same phenomenon. However, no firm proof that the two are causatively linked has been offered. Recently, a number of reports have shown that the rate of firming and rate of retrogradation are not the same.

The firming of bread crumb can be reversed by heating. This is one of the advantages of toasting bread; it is said to be "refreshened." The amylopectin crystal melts at about 60°C, but bread crumb continues to lose firmness as it is heated above 60°C to about 100°C (Fig. 25). This strongly argues that firming is not related to retrogradation of amylopectin.

A number of factors are known to alter the rate of staling or, perhaps more accurately, to produce softer bread that retains its softness over time. First, surfactants that complex with amylose are known for their ability to produce soft bread, presumably because starch, in the presence of surfactants, does not swell as much as starch alone. Second, inclusion of shortening in the bread formula alters the staling rate. Finally, the use of relatively heat-stable α-amylase in the bread formula retards staling. The temperature at which bread is stored also appears to be important, with higher temperatures retarding staling and cooler temperatures (above freezing) increasing the rate of firming.

Recently, a new model for firming has been suggested (Fig. 26). The model suggests that the gluten, the continuous phase of bread, is effectively cross-linked by the gelatinized starch in the bread. This suggested mechanism of firming appears to account for all the factors affecting firming.

Other Types of Leavened Products

SPECIALTY BREADS

The great preponderance of the bread produced in the United States is white pan bread. However, over the last decade, nonwhite or specialty bread has increased rapidly in popularity. The reasons for this are many and undoubtedly quite complicated. To name but a few, they are the natural food movement, the fiber fad, and the fashionable practice of maligning both the nutritional value and flavor of white pan bread. Whatever the reasons, they have led to much higher production of a greater variety of specialty breads in the United States.

White breads that are not baked in a pan are also specialty breads. Included in this type are hearth breads of the French, Italian, and Vienna types and a number of hard rolls. Other than the obvious difference of being baked free-standing on a hearth rather than supported in a pan, the difference between pan bread and these loaves as made in the United States is not large. In general, steam is used in the baking oven to produce a thicker crust. In some shops, the formula

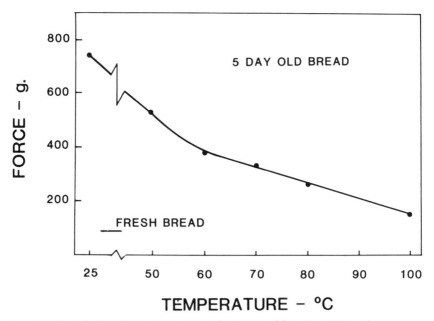

Fig. 25. Effect of reheating temperature on freshness of five-day-old bread.

is somewhat leaner (lower in fat, sugar, and milk) but not nearly as lean as the flour, water, salt, and yeast loaves of their European namesakes.

Wheat breads—breads made from 100% whole wheats, as well as combinations of whole wheat and white flour—are also increasingly popular types of breads. To be labeled as a whole-wheat loaf in the United States, the bread must be made from whole-wheat flour of specified granulation derived from the whole wheat kernel. Loaves of this type are small and compact, with a high density and low volume. Most wheat bread produced in the United States contains partly whole-wheat meal and partly a strong white flour. This results in a much better loaf volume and a lighter-tasting product. Many bakers also add small, or not so small, amounts of vital wheat gluten to such formulas to aid in improving the volume. The formula, other than the type of flour, does not vary greatly from that used for white pan bread, being rather rich in both sugar and fat.

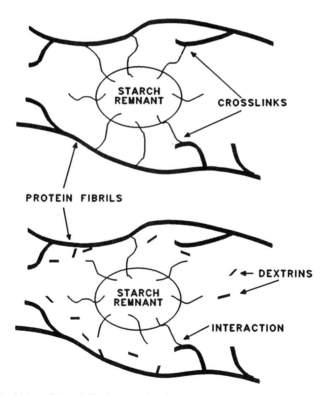

Fig. 26. Depiction of bread firming mechanism and antifirming role of dextrins.

In Europe, some bread is made from 100% rye flour; this type of bread is rarely seen in the United States. Most rye bread sold in the United States is made with white wheat flour with enough added rye flour to attain a desired flavor and appearance. Rye flour is generally produced as light or dark flour. Light rye is just a short patent flour, whereas the dark flour is more likely to be a "stuffed" straight grade. (The term *stuffed* means that the lower-grade flour stream is added to a straight-grade flour to produce a flour of lower grade.) The different flours are used to produce the color and flavor desired. The types of rye bread vary widely, from white wheat loaves containing only small amounts of rye to much darker and denser loaves. Rye flour does not have the type of gluten found in wheat and therefore produces a low-volume, dense loaf. The gas-retaining ability of rye flour appears to result from the pentosans that rye flour contains rather than from the proteins.

Two strange practices are associated with rye bread production in the United States. Formulas for most U.S. rye breads contain caraway seed, either whole or ground. It appears that U.S. consumers have been taught that rye bread should have a caraway flavor. This is unfortunate for consumers who like rye but do not care for caraway. An equally strange practice is the adding of caramel color to white bread that contains just enough rye to be so labeled and passing this off as dark rye bread.

Sourdough breads, both rye and wheat, are popular in certain areas of the United States. Before baker's yeast became commercially available, the sour system was all that was available as a leavening agent. Sours are mixtures of naturally occurring organisms that are found in flours. They are predominantly bacteria but also contain a number of wild yeasts. The yeast generally produces gas, and the bacteria are responsible for the acids. A number of bacteria also can and do produce gas.

Sours can be started by a number of techniques. Using old dough is probably the simplest. As a dough becomes old, the commercial yeast becomes inactive, as the pH is low. At the lower pH, bacteria are still quite active. With time, the food available for the organisms becomes limited; however, as new flour is added and the "starter" is fed or rebuilt, the bacteria become more predominant. One must guard against getting the pH too low, because below about pH 3.7, putrefaction bacteria take over and the odors are quite bad. Generally this is not a problem, as flour is a relatively good buffer. The desirability of a sour depends upon the flavor produced and the rate of gassing obtained. In general, the gas-producing ability of sours is low compared to that of commercial yeast. Consequently, proof time of sourdough bread is often long, of the order of several hours. Sourdough bread is often produced as round loaves and baked in ovens with steam to give a heavy crust.

FROZEN DOUGH

The use of frozen dough has increased in popularity over the last few years. The use of frozen doughs, particularly in in-store bakeries, has several advantages. For example, highly trained personnel are not required for the bake-off, and the store's space and equipment requirements are lower. In addition, the aroma of fresh baked bread in the supermarket is a definite advantage, and the frozen dough can be held until it is needed and then thawed and used, reducing waste. Other advantages include the production of large amounts of dough using trained personnel at a central location and the ease of shipping the relatively dense frozen doughs over wide distances. Disadvantages include shipping of the water in the dough and the need for good temperature control during shipping and storage.

The major disadvantage of frozen dough is the variable performance of the product after frozen storage. In general, doughs frozen with no fermentation perform better than do doughs frozen after fermentation. The products of fermentation appear to be detrimental to the yeast. After a relatively short fermentation time, doughs that are frozen take extended times to proof. In addition to the elimination of fermentation time, ascorbic acid is usually used as an oxidant in frozen doughs. For reasons that are not clear, ascorbic acid appears to give much better results than do other oxidants in frozen doughs.

A second problem with frozen doughs is the growth of ice crystals in the dough. As the ice crystals grow during storage, water that was hydrating the protein is removed. Upon thawing, this water does not go back into the protein but instead results in a wet, sticky dough with a coarse structure. Remixing or sheeting of the dough eliminates the problem, but this is usually not a viable option in a bake-off operation.

To maximize the effects both on yeast activity and dough rheology, the maintaining of a constant frozen temperature is beneficial. Perhaps a better statement is that variable temperature leads to increased damage. In summary, frozen dough with good performance after thawing can be produced by using ascorbic acid as an oxidant, eliminating fermentation, and storing the frozen dough at a constant temperature. It is also a general practice to add higher levels of yeast in frozen doughs. Whether this is a real advantage or not is not clear.

Often one sees discussion of the rate of freezing and thawing in regard to frozen doughs. A rapid rate of freezing is probably desirable, as it limits fermentation time. Bread dough, with its gas cells, is a reasonably good insulator; it is actually difficult to change either the freezing or thawing rate at any point except on the surface of the dough. Also, under normal conditions, the yeast probably does not freeze. The water around the yeast freezes, but the yeast cytoplasm probably just supercools and does not freeze until about $-35°C$. Commercially, dough is generally not cooled to that temperature. In laboratory experiments where the yeast was cooled to that temperature, the ability of the yeast to ferment was lost completely.

PUFF PASTRY AND OTHER SWEET DOUGHS

One could argue that most American bread is made from sweet dough, as the sugar content continues to creep toward 10%, based on the flour weight. However, this section concerns doughs with even higher sugar levels. Baker's yeast is fairly tolerant to osmosis but does ferment at a markedly slower rate at high sugar levels. Yeast contains very active invertase, and so sucrose is rapidly converted to glucose and fructose, which increases the osmotic pressure in the dough. Higher levels of yeast are generally used in sweet-dough preparations. Other than the sugar's effect on yeast, most sweet doughs do not present any particular problems. They tend to be sticky, particularly if they are warm. Keeping the dough temperature low is an advantage.

Puff pastry and Danish dough are made up of layers of dough and layers of shortening or butter. To obtain the layers, the dough is worked to a size, and then half the dough is covered with fat and the dough folded to sandwich the fat; the dough is then repeatedly sheeted and lapped. This gives rise to the many, many alternating layers of dough and fat. For the system to work correctly, the fat must have the correct physical properties (both softening point and rheological properties are important). Because the properties of the fat change rapidly with temperature, the dough temperature must be carefully controlled. A common practice is to sheet the dough a number of times and then return it to a retarder (refrigerator) to allow it to rest and cool before it is subjected to an additional series of sheetings.

Flour Quality for Breadmaking

To speak of flour quality without defining what product is being considered can be very misleading. A flour that is good quality for breadmaking may not be good quality for cookies. However, being good for one use may not automatically mean it is not good for another. In this section, the discussion of flour quality is confined to use for breadmaking.

In considering bread-making flour, we usually think of hard wheat with a relatively high protein content. However, in various part of the world, breads made from both soft and durum wheats can be found. Therefore, hardness does not appear to be an absolute requirement. The protein-content requirement does appear to be more absolute. It appears to be impossible to make a good-quality loaf of bread from flour that contains a low amount (8% or so) of protein. On the other hand, the protein content by itself does not assure a good quality. Therefore, both a certain quantity and quality of protein is needed in the wheat flour to produce a quality loaf.

The quantity of protein in a flour can be determined with good accuracy by several techniques. At this time, the quality cannot be determined so easily. In fact, our only reliable test for evaluating bread-making quality is a bread-making test. The choice of bread-making test is, of course, critical. It must be not limiting in any factor, so that the true ability of the flour to retain gas and give a large volume is fully expressed. A number of such tests exist in the literature.

Use of such a bread-making test and baking of a large number of samples provide data to plot loaf volume as a function of protein content (Fig. 17). An obvious correlation exists between higher protein content and higher loaf volumes. However, the correlation is not especially high (on the order of 0.80–0.85). If samples are separated on the basis of variety (cultivar), the correlation within a cultivar is much better. Shown in the plot in Fig. 17 is a family of regression lines. They are essentially straight lines above about 8% protein. The slope of the line is a measure of flour quality. Actually, the slope is the cubic centimeters of loaf volume per unit of protein and can be considered a measure of flour quality. A plot such as that shown in Fig. 17 is quite useful. It shows clearly the effect of both the quantity of protein and its quality (slope of the regression line). By baking wheat flours with a suitable bread-making procedure and knowing the protein content from such a plot, one can obtain a rapid estimate of the protein quality.

Another interesting aspect of the plot is that all the regression lines come together at about 8% protein. We cannot determine the protein quality of low-protein samples. Obviously, protein quality is not an important factor if one is dealing with low-protein flours. Another value of the plot is its usefulness in comparing samples of different protein contents. For two flours of different protein contents and producing loaves of different volumes (for example, flour A with 12.2% protein and a loaf volume of 958 cm^3 and flour B with 13.7% protein and a loaf volume of 1,075 cm^3), it is sometimes difficult to determine which has the better protein quality. The plot would show that flour A has the larger slope and thus the better quality. One always has the desire to simply divide the loaf volume by the protein content and assume that the larger number indicates the better quality. Such an approach, of course, assumes that a linear relationship of protein and loaf volume exists all the way from 0% protein. This is not true and can lead to erroneous results. Actually, a bread loaf made from 100 g of flour with no protein will have a loaf volume of about 400 cc.

As useful as the plot of loaf volume as a fraction of protein is, it does not prove the dependence of loaf volume on protein content and quality. It is only a statistical relationship. To prove that the protein is responsible for differences in loaf volume, good- and poor-quality flours must be fractionated into gluten, starch, and water-soluble fractions. These are then reconstituted into flours with baking properties equal to those of the original flours and into flours with the fractions interchanged. This experiment shows unequivocally that the gluten protein fraction is the fraction that controls the loaf volume and mixing requirement of flours.

REVIEW QUESTIONS

1. Why is wheat flour unique among the cereal flours in its ability to produce yeast-leavened products?
2. What is a minimum bread formula?
3. What is the role of yeast in breadmaking?
4. What is the role of salt in breadmaking?
5. What is the role of water in breadmaking?
6. What are the multiple roles of fat in breadmaking?
7. What is the role of sugar in breadmaking?
8. Why are bread flours malted?
9. What is keyholing?
10. What is commonly used as a bread softener?
11. What is commonly used as a dough strengthener?

12. What is the most common additive to control mold growth?
13. Outline a straight-dough process.
14. What are the advantages of the sponge-and-dough procedure?
15. Describe what happens during the mixing of a bread dough.
16. Why does a mixing curve produce a peak?
17. Why is it important to incorporate air into dough during mixing?
18. What is the relationship between the number of air bubbles in dough and the grain of the final bread?
19. What factors control the mixing time of a flour?
20. Explain how the pH of the dough affects mixing time.
21. Why does an overmixed dough lose its resistance to extension?
22. Why is bread fermentation an anaerobic process?
23. What are the purposes of punching or remixing a dough during fermentation?
24. How does yeast affect the rheological properties of dough?
25. Do the products of fermentation affect dough rheology?
26. Does a fully developed bread dough flow to fill the bread pan during proofing?
27. How much carbon dioxide is produced in a 100-g loaf during a 3-hr fermentation and a 1-hr proof period?
28. Is there sufficient CO_2 in a bread dough to account for the oven-spring expansion of the dough?
29. Does ethanol contribute to the volume of baked bread?
30. Explain how the crumb grain of bread can be controlled.
31. Explain how gas is retained in a bread dough.
32. Can "bread" of good volume be produced in a system containing no gluten?
33. Rye flour's ability to retain gas appears to be related to what fraction of the flour?
34. Explain how lipids plus starch can give "bread" of good volume.

35. What fraction of wheat flour appears to control the rate of gas diffusion in a bread dough?
36. Explain what happens when a dough is placed in a hot oven.
37. What temperature does the center of a loaf of bread attain during baking?
38. Explain how the size and shape of the crumb grain are related to the strength of the dough.
39. What is the difference between dough and bread?
40. Explain how dough is converted into bread.
41. What is responsible for crust staling?
42. What is retrogradation?
43. What controls the degree of staling?
44. How can bread with a thick crust be produced?
45. Explain how a starter works to produce a sourdough bread.
46. What are the advantages and disadvantages of frozen dough systems?
47. Why does the dough rheology change during frozen dough storage?
48. How are puff pastry and Danish made?
49. What is meant by protein content and protein quality when discussing bread flour quality?
50. What is the minimum amount of protein needed to produce a loaf of bread?
51. What is the loaf volume of bread made from 100 g of starch?

SUGGESTED READING

BARNES, P. J, ed. 1984. Lipids in Cereal Technology. Academic Press, Orlando, FL.

BLOKSMA, A. H., and BUSHUK, W. 1988. Rheology and chemistry of dough. Pages 131-217 in: Wheat: Chemistry and Technology, 3rd ed. Vol. II. Y. Pomeranz, ed. Am. Assoc. Cereal Chem., St. Paul, MN.

FINNEY, K. F. 1984. An optimized, straight-dough, bread-making method after 44 years. Cereal Chem. 61:20-27.

HOSENEY, R. C. 1984. Gas retention in bread doughs. Cereal Foods World 29:305-308.

HOSENEY, R. C. 1985. The mixing phenomenon. Cereal Foods World 30:453-457.

KULP, K. 1988. Bread industry and processes. Pages 371-406 in: Wheat: Chemistry and Technology, 3rd ed. Vol. II. Y. Pomeranz, ed. Am. Assoc. Cereal Chem., St. Paul, MN.

MacRITCHIE, F. 1980. Physicochemical aspects of some problems in wheat research. Pages 271-326 in: Advances in Cereal Science and Technology, Vol. 3. Y. Pomeranz, ed. Am. Assoc. Cereal Chem., St. Paul, MN.

MacRITCHIE, F., du CROS, D. L., and WRIGLEY, C. W. 1990. Flour polypeptides related to wheat quality. Pages 79-145 in: Advances in Cereal Science and Technology, Vol. 10. Y. Pomeranz, ed. Am. Assoc. Cereal Chem., St. Paul, MN.

MILLER, B. S, ed. 1981. Variety Breads in the United States. Am. Assoc. Cereal Chem., St. Paul, MN.

POMERANZ, Y. 1988. Composition and functionality of wheat flour components. Pages 219-370 in: Wheat: Chemistry and Technology, 3rd ed. Vol. II. Y. Pomeranz, ed. Am. Assoc. Cereal Chem., St. Paul, MN.

PYLER, E. J. 1988. Baking Science and Technology, 3rd ed. Vols. 1 and 2. Sosland Publishing Co., Merriam, KS.

SCHUSTER, G., and ADAMS, W. F. 1984. Emulsifiers as additives in bread and fine baked products. Pages 139-287 in: Advances in Cereal Science and Technology, Vol. 6. Y. Pomeranz, ed. Am. Assoc. Cereal Chem., St. Paul, MN.

SEIBEL, W., BRUEMMER, J. M., and STEPHAN, H. 1978. West German bread. Pages 415-456 in: Advances in Cereal Science and Technology, Vol. 2. Y. Pomeranz, ed. Am. Assoc. Cereal Chem., St. Paul, MN.

Soft Wheat Products

CHAPTER 13

The diversity of products made from soft wheat flour is truly amazing. Unlike hard wheat flour, which is used for one major product, bread, and numerous minor ones, soft wheat has more than one major use. The products made from soft wheat flour can be grouped into cookies, cakes, crackers, and pretzels. However, there is wide variation within each of these groups. Cookies not only vary in appearance and taste but also in the type of flour needed to produce a desirable product.

In addition to requiring soft wheat flour, most of these products are alike in being leavened chemically. Small amounts of yeast may be used in some pretzel formulas, but for the most part, the leavening is either by air or by the products of a chemical reaction.

Soft vs. Hard Wheat Flours

In Chapter 1, the structural differences between hard and soft wheats were discussed. These differences have significant effects upon the properties of the flours produced from these wheats and therefore upon their products. Hard wheats are, as their name implies, harder in nature. Thus, more work is required to reduce the wheat to a fine particle size. One result of that work is that a greater percentage of the starch is damaged during milling. Higher damaged-starch values are often viewed as a negative factor, particularly for cookie flours.

The factor that makes hard wheat hard also apparently has an effect upon the texture of the products made from that flour. For example, cookies made from hard wheat flour are almost invariably hard in texture. The effects of the various flour components on product quality are discussed in greater detail later in this chapter.

Chemical Leavening

Four gases are used for leavening. They are carbon dioxide, water and/or ethanol vapor, ammonia, and air. Air, of course, is really a mixture of gases and is present (and therefore used) in all products. Water is also present in all baked products, but its leavening effect is quite limited in most products because it has a relatively high boiling point. Water vapor is an effective leavener only when the product is heated at a very fast rate, as is done to make saltine crackers. Yeast, the major leavener for bread, produces carbon dioxide and ethanol. In general, yeast is rarely used in soft wheat products. As discussed in Chapter 10, it has a strong effect on dough rheology. This effect is generally not desirable in soft wheat products, which may account for the nonuse of yeast in them. Carbon dioxide can also be produced by chemical reactions of bicarbonate or carbonate with acids. These chemical reactions are most commonly used for leavening in soft wheat products.

The most common sources of carbon dioxide are sodium and ammonium bicarbonates. When heated, *ammonium bicarbonate* breaks up to give three gases, as shown below:

$$NH_4HCO_3 \xrightarrow{\Delta} NH_3 + CO_2 + H_2O$$

Ammonium bicarbonate can be used only in products that are to be baked to a low moisture content (~5%). If the product retains more than a few percent water, it also retains ammonia. Even a small amount of ammonia in the product makes it inedible. Therefore, ammonium bicarbonate has limited usefulness. It is used fairly widely in dry cookies and in some snack cracker types of products. It does have the advantage that it leaves no residual salt after it reacts. Residual salts could affect the flavor or the dough rheology or both.

Potassium bicarbonate is also a potential source of carbon dioxide for leavening. However, it is generally not used because it tends to be hygroscopic and to impart a slight bitterness to the products.

The most popular leavening agent by far is *sodium bicarbonate* (baking soda). Its popularity is based upon a number of advantages it offers. For example, 1) it has a relatively low cost, 2) it is nontoxic, 3) it is easy to handle, 4) it gives a relatively tasteless end product, and 5) the commercially produced product is of high purity. *Sodium carbonate* could also be used as a source of carbon dioxide but is not used. Its major disadvantage is its high alkalinity, which increases the danger of getting a localized region of such high pH that it might be detrimental to the product.

To understand the use of carbon dioxide as a leavening agent, one must first understand a little of the chemistry of carbon dioxide. It reacts with water to form carbonic acid:

$$CO_2 + H_2O \rightarrow H_2CO_3$$

Thus, carbon dioxide can exist as the free CO_2 or as one of two ion species, HCO_3^- or $CO_3^=$. The relative proportion of each is determined by the pH and temperature of the solution. The effect of pH in determining which species can exist is shown in Fig. 1. No leavening gas (CO_2) is available if the pH stays above 8.0. Many soft wheat products end up with a pH near 7.0, where only a part of the CO_2 is in the gaseous state.

To obtain a larger yield of gas and to control the rate of carbon dioxide evolution, acids are added to doughs. Sodium bicarbonate is quite soluble in water and dissolves rapidly when the dough or batter is mixed. That, of course, raises the pH of the batter or dough to the point where no carbon dioxide is released. To obtain significant amounts of gas, the dough or batter must contain an acid. Many ingredients used in baking are sources of acid. Acidic fruits or buttermilk are obvious examples. If no acid is available natively, one must be added to the formula. Heating of $NaHCO_3$ in an aqueous system causes a disproportionation, with about half of the CO_2 being released and the remainder going to sodium carbonate.

If the formula does not contain the acid, then we must use a combination of baking soda and an acid (i.e., *baking powder*). Baking

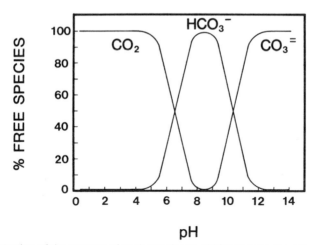

Fig. 1. Illustration of the percent of each species vs pH for carbon dioxide.

powders consist of mixtures of baking soda, one or more acid salts, and a diluent. By law, a baking powder must yield not less than 12% available carbon dioxide. That regulation effectively establishes the level of soda. The acid or acids used are determined by their neutralization values. The inert diluent is usually dried starch. Its primary function is to physically separate the soda and acid particles and prevent their premature reaction.

Baking powders are either single- or double-acting. A double-acting baking powder is one that contains two acids, one that reacts (becomes soluble) at room temperature and another that reacts when the product is heated.

The amount of acid required in a formulation depends upon the amount of soda and the *neutralization value* of the acid. Because the acids used are acidic salts, the stoichiometry of their reactions is often not clear. Therefore, the concept of neutralization value was developed.

$$\text{Neutralization value} = \frac{\text{g of NaHCO}_3 \times 100}{100 \text{ g of acidic salt}}$$

Generally the pH of the product should not be affected by the leavening reaction. However, if the correct amount of acid is not used, the properties and taste of the product will change. For example, an excess of soda generally gives the product a soapy flavor. The color of many products is highly dependent upon the pH.

Several leavening acids are used in the baking industry. In general, the acids vary in their rate of reaction at various temperatures (Fig. 2). The properties of the most common leavening acids are given in Table I.

Cream of tartar (monopotassium salt of tartaric acid), the original leavening acid, was obtained as a by-product from wine production. It reacts readily at room temperature. Because it is relatively expensive, it has been largely replaced by *monocalcium phosphate* in most applications. Monocalcium phosphate also reacts readily at room temperature and is widely used as the fast-acting component in double-acting baking powders.

A number of *sodium acid pyrophosphates* (SAPPs) are on the market. They vary in their reaction rates, depending upon how they are made. SAPPs are used widely in canned biscuits and in cake doughnuts, both of which have unique leavening requirements that are handled only by the SAPPs. The major problem with SAPPs is the "aftertaste" that they leave in the mouth and on the teeth. This so-called "pyro" taste is quite noticeable in these products. It apparently comes from the exchange of calcium from the teeth for the sodium in the disodium phosphate that results from the leavening reaction and is the result of the enzyme action that splits the pyrophosphate. Attempts to limit the effect of the disodium phosphate by adding various forms of calcium to the formula have been only partially successful.

Sodium aluminum phosphate (SALP) is the newest of the leavening acids. It is widely used as the second (higher temperature) acid in double-

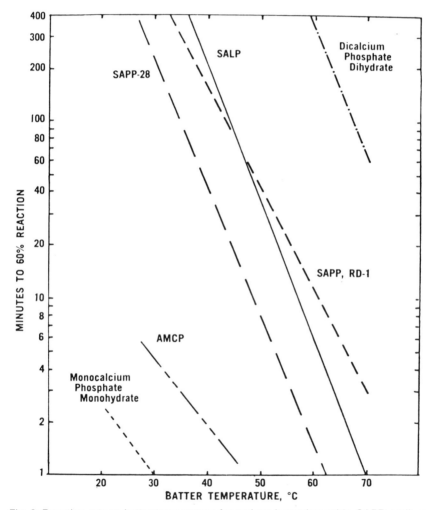

Fig. 2. Reaction rate vs batter temperature for various leavening acids. SAPP, sodium acid pyrophosphate; SALP, sodium aluminum phosphate; AMCP, anhydrous monocalcium phosphate; RD-1 is a grade of SAAP.

acting baking powders and also in commercial baking mixtures. It not only does a good job of leavening but also gives strong products with a strong crumb texture.

Sodium aluminum sulfate (SAS) was the most common second acid in baking powders before SALP became available. It is still used in some formulations. The major problems with SAS are its weakening effect upon crumb texture and its slightly astringent taste.

Dicalcium phosphate is not an acidic salt and thus would not be expected to give a leavening reaction. However, at higher temperatures, the salt disproportionates and does give an acidic reaction. Generally, this happens at too high a temperature for the salt to be useful as a leavening acid, but it is useful for adjusting the final pH of the product.

Glucono-δ-lactone is an internal lactone that produces an acid as it is hydrolyzed. Its usefulness in baked products is somewhat limited, as the hydrolysis occurs over a wide temperature range. It also tends to produce products with a slightly bitter aftertaste. Its major advantage is that it does not produce the normal salts found with other leavening acids. However, it tends to be expensive compared with other leavening acids.

Besides their obvious effects upon the amount and rate of gas production and, in some cases, their effect upon the product's taste, the salts produced by the leavening reaction can affect the rheology of the product. In general, di- or trivalent ions tend to increase the

TABLE I
Properties of Common Leavening Acids

Acid	Formula	Neutralization Value	Relative Reaction Rates[a]
Cream of tartar (monopotassium tartrate)	$KHC_4H_4O_6$	45	1
Monocalcium phosphate monohydrate	$CaH_4(PO_4) \cdot H_2O$	80	1
Anhydrous monocalcium phosphate	$CaH(PO_4)$	83.5	2
Sodium acid pyrophosphate	$Na_2H_2P_2O_7$	72	3
Sodium aluminum phosphate	$NaH_{14}Al_3(PO_4)_8 \cdot 4H_2O$	100	4
Sodium aluminum sulfate	$Al_2(SO_4)_3 \cdot Na_2SO_4$	100	4
Dicalcium phosphate dihydrate	$CaHPO_4 \cdot 2H_2O$	33	5[b]
Glucono-δ-lactone	$C_6H_{10}O_6$	50	...[c]

[a] Relative rate; 1 = reactive at room temperature, 5 = requiring oven temperature for reaction.
[b] Generally reacts too slowly to be a leavening acid; used to adjust final pH.
[c] Reaction rate depends on many factors in addition to temperature.

elasticity of the product, and sulfate ions tend to decrease its elasticity. These ions presumably act by forming cross-links with the proteins in the batters.

Cookies

In general, cookies are products made from soft wheat. They are characterized by a formula high in sugar and shortening and relatively low in water. Similar products made in Europe and the United Kingdom are called "biscuits." The biscuits made in the United States are more accurately defined as chemically leavened bread. The diversity of cookie products is quite wide; they vary not only in formula but also in type of manufacture. In addition, a number of products do not fit our definition of cookies but are still called cookies mainly because they don't fit elsewhere.

In commercial practice, cookies are baked in long tunnel ovens. Typically, the baking zone of such an oven is about 1 m wide and 100–150 m long. The cookies are generally baked on a solid steel band that conveys the product through the oven at a rate that will produce the desired bake time (Fig. 3). Perhaps the best way to classify cookies is by the way the dough is placed on the baking band. Such a classification allows us to divide cookies into four general types.

ROTARY-MOLD COOKIES

For this type of cookie, the dough is forced into molds on a rotating roll. As the roll completes a half turn, the dough is extracted from

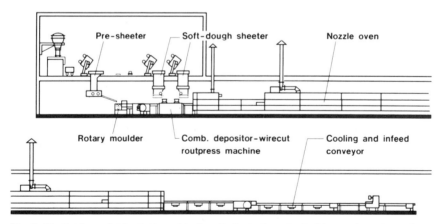

Fig. 3. Illustration of a typical cookie line that can be adapted to produce a number of types of cookies.

the cavity and placed on the band for baking (Fig. 4). The consistency of the dough must be such that it will feed into the cavity with no voids but still be extracted from the cavity without being distorted. During baking, the cookie should neither rise nor spread. Any movement will distort the design embossed on the cookie.

Formulations for rotary-mold cookies (Table II) are characterized by fairly high sugar and shortening levels and very low amounts of water (<20% based on the flour and including the moisture in the flour). The typical dough is crumbly, lumpy, and stiff, with virtually no elasticity. The gluten in the dough should not develop during mixing. Much of the cohesiveness of this type of dough comes from the plastic shortening used. The cookie does not spread during baking because of its low water content.

Rotary-mold cookies are economical to produce. A small amount of water is added to the dough and therefore less energy is required to remove it during baking. A rotary mold is generally used if it can make the desired product. Typical examples of cookies made on rotary-mold equipment are Oreo and Hydrox cookies.

Fig. 4. Rotary mold operation.

CUTTING-MACHINE COOKIES

The process for this type of cookie includes both rotary-cut and stamped doughs. The dough is made into a continuous sheet and the product is cut out from it. Typical examples are animal cookies, gingerbread men, etc. (Fig. 5). The formula contains much more water than do formulas used with a rotary mold. The sugar content is rela-

TABLE II
Typical Formula for Rotary-Mold Cookies

Ingredient	Parts[a]
Flour (soft wheat)	100
Sugar	20
Shortening	25
Acid cream (cream of tartar)	0.5
Sodium bicarbonate	0.5
Salt	1.5
Condensed milk	6.0
Whole egg	3.5
Butter	1.2
Lecithin	0.3
Malt	1.4
Water (variable)	about 10.0

[a] Based on flour.

Fig. 5. Rotary-cut pieces from a sheeted dough with the scrap being lifted away.

tively low compared to the level in most cookies. Because a sheet of dough is made, the gluten is developed in this system. The gluten development stops spread and distortion during baking of this type of cookie.

WIRE-CUT COOKIES

A relatively soft dough is extruded through an orifice and cut to size, usually by a reciprocating wire. The dough must be cohesive enough to hold together, yet short (noncohesive) enough to separate cleanly when cut by the wire. A typical formulation, based on flour weight, may contain 50–75% sugar, 50–60% shortening, and up to 15% eggs. Wire-cut mechanisms can handle a range of cookie doughs, and a wide variety of products can be made. Wire-cut cookies rise and spread as they are baked. The final size of the cookie is determined by the formula and the flour used. In addition to such common cookies as chocolate chip and oatmeal, fig or date bars can be coextruded and wire-cut.

SUGAR WAFERS

This type does not really fit our definition of cookies, but it does not fit elsewhere. Ice-cream cones and sugar wafers differ from other cookies. The formula (Table III) contains no sugar, essentially no fat, and an excess of water.

The flour is usually of a short extraction and is generally from a white wheat. The bran specks from red wheat would show up in the product. If the flour is too weak (has no gluten strength), excessively dense products generally result. If the flour is too strong (has excessive gluten development), the wafers are hard and flinty. The ingredients are generally mixed to produce a lump-free batter. Overmixing, or for that matter, simply holding the batter too long, causes the gluten strands to separate from the aqueous mixture. Therefore, a common practice

TABLE III
Typical Formula for Wafers or Ice Cream Cones

Ingredient	Parts[a]
Flour	100
Water	135
Sodium bicarbonate	0.375
Salt	0.5
Lecithin	1.5
Coconut oil	1

[a] Based on flour.

is to make small batches and use each batch in a relatively short time.

Even though sodium bicarbonate is in the formula, its major function is to adjust the pH, not to act as a leavening agent. The primary leavening in this product is the steam produced from water. Baking is usually done in a closed-plate system that resembles waffle irons (Fig. 6). Sets of plates travel on an endless chain through the heating chamber. The plates open automatically and are filled with batter. A major problem is products that stick to the plate and go through another cycle. The double shot of batter often springs the plates so that they no longer operate correctly. Continuous ovens have been built and are being used.

After baking, the product is cut into "books" and the filling placed between the various layers. The filling generally consists of fat, sugar, flavors, and ground scrap product. The breakage is quite high for this type of product, particularly for ice cream cones.

Cookie Flour Quality

As with all quality determination, we must first define what constitutes good quality. In general, cookie quality can be summarized

Fig. 6. Sugar wafer baking unit, showing the plates.

in two general terms. First is the size of the cookie, both the width and the height. Second is how the cookie bites; good quality cookies must have a tender bite.

The importance of cookie size can be appreciated if one considers that the cookie box, with its appropriate labeling, including net weight, is ordered months before the cookie is baked. If the cookie spreads too much, it cannot be put in the box without breaking. If the spread is too little, then the box is not completely filled and the net weight is in error. The problem is avoided with rotary-mold or rotary-cut cookies, which have no spread, but all types of cookies cannot be produced on that equipment. Another way to avoid the problem is to sell cookies in a bag filled to a certain weight. However, this is generally not a satisfactory approach for quality cookies.

The nice tender bite that we associate with good quality cookies comes from two major factors. The use of fat, or "shortening," as it is usually called, produces a short product. The word *shortening* came from this usage. The second factor is the flour. Generally, a good soft wheat flour gives products with a tender bite. The chemistry responsible for this difference between hard and soft wheats is not clearly understood. Perhaps the factor that makes hard wheat hard also makes cookies from hard wheats hard.

What Happens During Cookie Baking

The primary function of the mixing step in making cookie dough is to produce a uniform mixture and to incorporate air into it. For high quality products, that often requires creaming (mixing the sugar and shortening) as a preliminary step. The advantage of creaming is discussed later in this chapter in regard to cakes. In general, we do not want to develop the gluten during mixing. Development of the gluten leads to tougher products and generally to cookies that will not spread. A small amount of gluten development is needed in dough that will be sheeted and cut.

In general, the retarding of gluten development is not much of a problem. With the high level of sugar in most cookie formulas and the relatively high pH because of the sodium bicarbonate, the gluten proteins do not hydrate very readily. Gluten cannot develop if it is not hydrated.

When cookie dough is heated in an oven, a number of events occur. They can be shown by heating a cookie dough in a differential scanning calorimeter (DSC) (Fig. 7). First, the shortening in the dough melts. Shortening helps give the dough part of its plastic character, so the dough containing melted shortening is more free to flow under the force of gravity. The second thermal event is dissolution of the sugar. During mixing, only about half the sugar is dissolved; the remainder stays in a crystalline form until the dough is heated. Then that sugar dissolves.

When sugar dissolves, it increases the volume of solution in the system. Each gram of sugar, when dissolved in 1 g of water, produces 1.6 cm^3 of solution. One of the effects of this can be seen in Fig. 8, where the addition of dry sugar turns a powder system into a suspension. The increase in total solution also has a pronounced effect on the cookie dough. The increased solvent makes it sticky. Thus, if we attempt to produce cookies from syrup rather than crystalline sucrose, we get sticky doughs that do not machine well.

Continued heating of the cookie dough in the DSC gives an additional endotherm, presumably starch gelatinization at about 115°C. Because the moist cookie dough does not reach this temperature during baking,

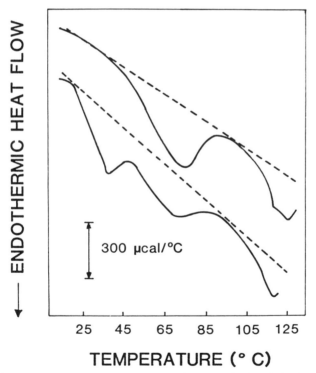

Fig. 7. Differential scanning calorimetry curve for a cookie dough (top) and a dough containing no shortening (bottom). The dashed lines are the baseline.

we must assume that the starch does not gelatinize during cookie baking. This is confirmed by grinding the dried cookie crumb and running it in the DSC. An endotherm equal in size to that found in the unbaked cookie dough is found, which shows that starch is not gelatinized during cookie baking. However, there are many cookie formulas, and those low in sugar may well have some starch gelatinization.

In summary then, as the cookie dough enters the oven and starts to heat, the shortening melts, giving the dough more fluidity. At the same time, the sugar is dissolving and thereby increasing the solution (water plus sugar). This also increases fluidity and allows the dough to spread as a function of gravity. The leavening system becomes active and expands the piece in all directions. The cookie continues to spread until the apparent viscosity of the system becomes too large. Because starch is not gelatinized, the viscosity is presumably a property of the flour proteins.

Certain cookies, gingersnaps for example, have a cracking pattern on the surface that develops during baking. The phenomena has been

Fig. 8. Starch-H_2O mixture (1:1, left) and starch-H_2O-sugar mixture (1:1:1, right).

explained as follows. Moisture is lost from the surface of the cookie at a rapid rate during baking. The hot air in the oven has the capacity to hold a large amount of water. As the water from the surface is lost, it is replaced by water diffusing from the interior of the cookie, and the sugar, not being volatile, is concentrated. Sucrose is, of course, the most popular sugar used in cookies. One of the unique properties of sucrose is its tendency to crystallize. In fact, it will crystallize at the surface of the cookie during baking. After the sugar has crystallized, it no longer holds the water that gives a moist and moldable surface. Thus, the surface dries and breaks as the leavening system expands the cookie, producing the cracked surface. Small amounts of high-fructose corn syrup or certain other sugars interfere with sucrose crystallizing and thus destroy the cracking ability (Fig. 9). A casual examination of gingersnaps on the market will tell you which contain corn syrups and which do not.

Essentially all cookies, except those that are dried to a very low moisture content, are soft and quite flexible when they come out of the oven. With time, they become firm and often brittle. Some, such

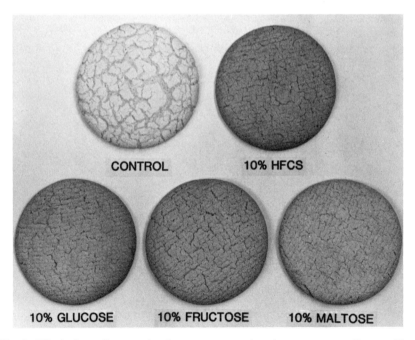

Fig. 9. Effect of small amounts of corn syrup and various sugars on the cracking pattern. HFCS, high-fructose corn syrup.

as sugar-snap or gingersnap cookies, give an audible snap when they are broken. This appears to be because of the crystallization of sucrose. When the sugar dissolves, the syrup gives the cookie flexibility. Indeed, if we want to maintain a soft cookie, we need only to add a sugar or other substance that interferes with sucrose recrystallization. If sucrose is allowed to crystallize, the water that was associated with the sugar is no longer controlled by the sugar and is free to migrate to other components. Because the cookie is low in moisture after baking (2–5%), many of its components (the starch or protein, for example) are glassy and therefore brittle. It is the brittle break that causes the snap in the cookie. The change in texture (measured by the force necessary to fracture the cookie) with time has been shown with force deformation curves (Fig. 10).

Recently, the cookie industry has produced a number of new products. This is the result of new companies entering the business and of the new dual-texture cookie. A dual-texture cookie can be made by using two doughs that are kept separate. One contains sucrose and the other a crystallization inhibitor (corn syrup).

Crackers

As with the case for cookies, our definition of crackers must be quite broad, as there are many types of crackers. In general, crackers contain

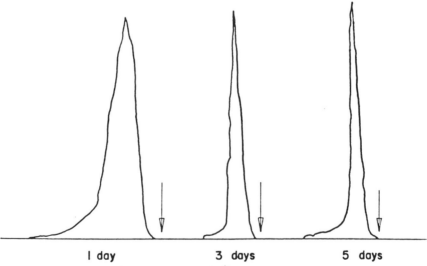

Fig. 10. Instron curves for compression of sugar-snap cookies, showing effect of time on curve area. Arrows indicate starting points.

little or no sugar but moderate to high levels of fat (10-20%), based on flour weight. The doughs generally contain low levels of water (20-30%). The leavening agent is either water vapor or a chemical leavening.

SALTINES

Saltine crackers are distinguished by their long fermentation time and their particularly light and flaky texture. They are made by a sponge and dough process with a formula similar to that given in Table IV. The sponge fermentation is generally about 16 hr. During this fermentation, the pH drops from about 6.0 to about 4.0. An inoculum is generally used to obtain this pH drop. The inoculum is called a "buffer" in the industry and is generally an old sponge. A general practice in the industry is to wash the dough troughs only on weekends. Thus, as the week progresses, less buffer is required, as the trough itself acts as an inoculum. The quality of saltine crackers can vary widely, depending upon the activity and amount of buffer used.

When first viewing a cracker sponge formula, many people want to increase the level of yeast and reduce the sponge fermentation time. This, however, assumes that the yeast fermentation is the important event of saltine cracker fermentation, which is not correct. Yeasts are always contaminated with bacteria. Apparently, it is the bacteria that play an important role in the changes occurring during fermentation. Flour contains only limited amounts of carbohydrates that can be utilized by either the yeast or the bacteria. Thus, when we "set" (mix) a sponge, the yeast and the bacteria compete for the fermentables. By using low levels of yeast and the added buffer (inoculum), we give the bacteria the upper hand in that competition. If we increase the amount of yeast, then the yeast will win the battle for the fermentables.

TABLE IV
Typical Saltine Cracker Formula[a,b]

Ingredients	Sponge (%)	Dough (%)
Flour	65.0	35.0
Water	25.0	...
Yeast	0.4	...
Lard	...	11.0
Salt	...	1.8
Soda	...	0.45

[a] Ingredients based on flour weight.
[b] A "buffer" is often added to the sponge to inoculate the system.

During fermentation, the dough is modified so as to become less elastic. Flour contains a proteolytic enzyme that has an optimum pH of about 4.1. The action of this enzyme is thought to be important in modifying the dough's texture. This may also be the reason that the sponge must reach a pH of 4.1.

After its fermentation, the sponge is mixed with the other dough ingredients and the dough flour. Included in the dough ingredients is sufficient sodium bicarbonate to bring the dough pH to above pH 7.0. For the correct taste and texture, the baked saltine should have a pH of about 7.2. Generally about 0.1 of a pH unit is lost during baking.

The dough fermentation period is usually about 6 hr. Because of the high pH, the bacteria are not active. Thus, we would expect the yeast fermentation to be predominant. Except for any changes brought about by the yeast, the dough is allowed to relax.

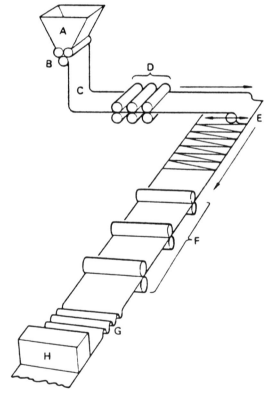

Fig. 11. Processing of a cracker dough. A, Dough hopper; B, forming roll; C, dough web; D, reduction rolls; E, lapper; F, final reduction; G, relaxing curl; H, cutter-docker.

SOFT WHEAT PRODUCTS / 293

Thus, 24 hr after the sponge was set, the dough is ready to be processed (Fig. 11). The dough is layered (laminated or lapped, Figs. 12 and 13) into eight layers of about 0.3 cm each. After the layering, the dough is passed through rolls to reduce the thickness from 2.5 cm to about 0.30 mm. This is done in a number of steps and with a 90° "turn" involved. In this way, the dough is sheeted in both directions and is

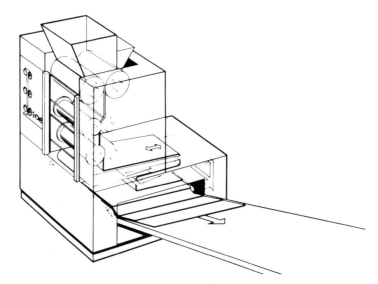

Fig. 12. A see-through view of forming rolls and laminator for a cracker line.

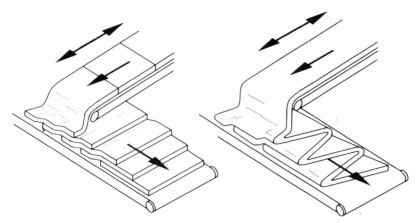

Fig. 13. Illustration of cutting lapping (left) and reciprocating lapping (right) of cracker doughs.

strong in both directions. After sheeting, the dough is cut and *docked* (Fig. 14). Cutting is usually done with a rotary cutter, which cuts the individual crackers to size but leaves them together in a continuous sheet. After cutting, the *docking* is done with docker pins that have a blunt end of about 0.15 cm in diameter. The purpose of docking is to pin the dough together (attach the top and bottom surfaces together) so that it won't separate into layers. After docking, the dough is salted (~2.5% of dough weight) and then baked. Typically, the oven's baking chambers are 100 m long and about 1 m wide. The baking surface is a mesh band so moisture can be lost from the bottom of the cracker; otherwise the cracker will cup (curl) during baking. Obviously, this is undesirable for a number of reasons.

Baking time is usually about 2.5 min at about 230°C. The rapid heating vaporizes water while it is still inside the dough and thus puffs the cracker (Fig. 15). Heating more slowly would just lead to loss of the water at the surface and no puffing. After baking, the crackers are cooled carefully (slowly) so that they will not *check* (develop tiny cracks that lead to breakage upon shipping). The sheets of crackers are then mechanically broken into the individual crackers and packaged. Much of the texture and desirability of saltine crackers is because of their low moisture content. Generally, the crackers contain about 2% moisture fresh from the oven. They are packaged to maintain that low moisture, as even small increases in moisture decrease the desirability of the product. Notice the layers of packing material designed to accomplish this the next time you open a package.

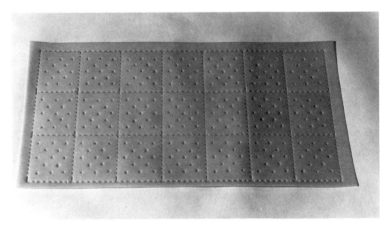

Fig. 14. Cracker dough after cutting and docking.

SNACK CRACKERS

Snack crackers vary much more widely in formula than do saltines. In general, they contain more shortening and much higher levels of flavoring materials. They generally do not contain yeast and are not given an extended fermentation period. They are chemically leavened. The doughs are mixed once with all the ingredients, allowed to rest, sheeted and laminated, and then cut and docked. Cutting is usually in a more unique shape than found with saltines, and this gives rise to more scrap dough that must be added back to the doughs. In general, the texture of snack crackers is denser than that of saltines.

Fig. 15. Baked crackers with a cross section showing the internal structure.

Layer Cakes

Cakes, like cookies, are characterized by a high level of sugar in the formula. The difference between the two is that cakes also contain relatively high levels of water. Because the molar sugar concentration is much lower in cakes than in cookies, the starch gelatinizes during baking. Therefore, cakes set when baked, giving a light product. The set is caused, at least in part, by starch gelatinization. In cookies, the starch does not gelatinize and therefore the structure collapses.

Because the light structure of cakes is important, we must be concerned about how to obtain that structure. As was discussed in Chapter 12, no new cells can be created either by chemical leavening or by yeast. Thus, air must be incorporated into the batter in the form of small air cells (nuclei) during mixing. For bread, the number of cells can be increased with with punches or remixes. However, for cakes, all the cells must be created during mixing. We can and do lose cells later in the process, but we cannot create any new ones.

Cakes can be divided into three different types, depending upon how the air is incorporated into the batter. Much of the confusion found in the literature on cakes is caused by not clearly defining what type of cake is being discussed. The first type of cake is made with *multistage mixing*. This is the classic procedure that starts with a creaming step. Fat and sugar are mixed together to form a cream. The purpose of the creaming step is to incorporate air into the fat. Two and even three subsequent mixing steps then incorporate the liquids and flour to form the final batter. Creaming steps are also used for certain types of cookies. Creaming offers a number of advantages; for example, a large number of air cells can be formed, which leads to a fine texture. Also, the batter can sit for extended periods of time because the air is in the fat, where it is stable (immobile). As the cake batter is heated and the shortening melts, the air cells are released into the aqueous phase, where the leavening gases can diffuse into them and thereby leaven the cake.

The second type of cake is made with *single-stage mixing*. This type is bought as a box mix; you just add the liquids and mix. In this case, the air is incorporated directly into the aqueous phase. This can be accomplished because the mix contains surfactants that lower the interfacial tension and allow the air to be incorporated directly. Propylene glycol monostearate is a common surfactant used for this purpose (see Fig. 3 of Chapter 12).

In the third type of cake, the air is incorporated directly into the aqueous phase by *mechanical means* rather than by using surfactants.

This is the type that is made commercially, using a high-speed mixing machine.

The cakes' properties also vary, depending upon the procedure used. For example, cakes made by the creaming procedure generally have a very fine grain. Those made by single-stage mixing (box cakes) are generally quite delicate and not suitable for shipping. Therefore, they are not suitable for commercial cakes, which must be handled and abused a great deal. When air is incorporated directly into the aqueous phase, the batter is not as stable as is a creamed batter.

Gas can diffuse in a batter and, because the pressure in a small bubble is larger than the pressure in a large bubble (see Chapter 12), there is a tendency for the small bubbles to disappear and the large bubbles to become larger (i.e., large bubbles eat little bubbles). This phenomenon is controlled by diffusion: gas diffuses out of the small bubbles into the aqueous phase and then into the larger bubbles. Because of their greater buoyancy, the larger bubbles overcome the high viscosity in the batter, rise to the surface, and are lost. Thus, such cake batters should not be allowed to sit for long periods of time before being baked.

In making cake mixes of the single-stage mixing type, the mix must be run through a *cake finisher*. This is essentially a grinder. Apparently, the purpose of the finishing is to bind the shortening to the flour. Microscopic examination of the finished mix shows no free fat in the mix, even though it contains 20–25% fat. Presumably, free fat would destabilize the aqueous foam that is produced as a result of mixing.

The discussion to this point has centered on layer cakes. A typical formula for white layer cake is given in Table V. This is a *high-ratio* cake; it contains more sugar than flour. Besides sugar's obvious function as a sweetening agent, it also has a tenderizing effect on the crumb,

TABLE V
Typical Formulas for Three Types of Cakes

Ingredients	Rich White Layer Cake	Angel Food Cake	Commercial Pound Cake
Flour	100	100	100
Sugar	140	500	100
Shortening	55	...	50
Eggs			
Whites (fresh)	76	500	...
Whole (fresh)	...	...	50
Milk (fresh)	95	...	50
Baking powder	1.3	...	...
Cream of tartar	...	20	...
Salt	0.7	...	...

presumably from the delayed gelatinization of starch. Reducing sugars are often added to the formula in the form of milk (lactose) or fresh egg whites (glucose) to give browning. Sucrose is nonreducing and, because the formula contains no yeast (and therefore no invertase to hydrolyze sucrose to reducing sugars), it does not give browning. Also, because the pH of the system is basic, no chemical hydrolysis of sucrose to glucose and fructose occurs. Therefore, when sucrose is the only sugar in the formula, no browning takes place and the cake is quite white.

Other Types of Cakes

EFFECT OF FORMULA INGREDIENTS

Shortening. The effect of emulsified shortening (mono- and diglycerides) and temperature on the viscosity of cake batters made with the AACC white layer cake formula are shown in Table VI. Batter viscosity at ambient temperature increased with increased levels of shortening in the batter. However, the minimum viscosity of the batters when heated decreased with increased shortening levels. This suggests that, while the solid (plastic) shortening increases the viscosity, it may decrease viscosity in the oven after it melts and creates a liquid phase. The viscosity of a cake batter is very important during the baking step. Too low a viscosity allows large bubbles to rise to the surface and escape. It also allows starch granules to accumulate at the bottom of the cake pan and produce a rubbery layer during baking.

The level of shortening in the cake formula does not affect the onset of starch gelatinization. However, high levels of shortening decrease the slope (rate) of the increase in the viscosity-versus-temperature curve. This suggests that the shortening decreases the rate of starch swelling.

TABLE VI
Effect of Increasing Levels of Emulsified Shortening on AACC White Layer Cake Batters

Shortening (% fwb)	Viscosity (cP)		Onset Temperature (°C)	Onset Slope (cP/°C)	Cake Volume (cm³)
	Ambient Temperature	Heated Batter[a]			
30	1,540 ± 20	108 ± 5	83 ± 0.5	81	995 ± 10
50[b]	1,950 ± 25	85 ± 5	83 ± 0.5	56	955 ± 10
80	2,250 ± 25	50 ± 5	83 ± 0.5	43	845 ± 5

[a] At the minimum point.
[b] Control.

Surface-Active Agents. The addition of certain surfactants (i.e., propylene glycol monostearate) to the formula of high-ratio, single-stage cake mixes is essential for incorporating sufficient air into the batter during mixing. The addition of surfactants to such a formula increases the batter's viscosity. Presumably, this is a reflection of the incorporation of additional air into the batter. The incorporation of air is shown as well by the decreased specific gravity of the batter containing the surfactants. The batter maintains this higher viscosity throughout the heating period. The presence of surfactants also changes (decreases) the rate of increase in viscosity (i.e., the slope of the increase in viscosity-versus-temperature curves). In this regard, the surfactant mimics the effect of shortening.

Baking Powder. It is also important to pick the correct baking powder for the cake. Generally, a double-acting powder is used. The first acid acts at room temperature to help nucleate the batter; the second acts during the oven stage. The leavening action must be timed so that the gas is produced while the batter can still expand, but not so early that a major part of the gas may diffuse out. However, if the gas is released too late, not only can the batter not expand, but the cake's grain can be destroyed by the excess pressure that is developed. Besides the obvious effects in releasing gas, the choice of leavening acids can also have other effects. Di- and trivalent cations such as calcium (+2) and aluminum (+3) tend to impart a greater resiliency to the cake crumb. On the other hand, sulfate ions tend to weaken the crumb. Also, different leavening acids tend to buffer at slightly different pHs. For example, $NaHPO_4$ buffers at 7.3, whereas aluminum phosphate does so as low as 7.1. This can change the appearance of the crumb. A change of only 0.2 of a pH unit has quite a noticeable effect on the whiteness of the cake crumb.

Cake Flour. Flours for high-ratio cakes must be treated with chlorine gas (Cl_2). Use of flours that are not chlorine-treated gives cakes that collapse in the oven. The literature shows that chlorine reacts with almost all flour constituents. The rate of reaction is quite rapid with flour and much slower with starch. However, fractionation and reconstitution studies have shown that the reaction with the starch is responsible for the improvement in baking properties. The amount of chlorine gas used is critical. When chlorine reacts with organic material, hydrochloric acid is usually produced. Thus, the pH of the flour is, generally, a good measure of the amount of reaction and therefore of the amount of chlorine added. Generally a pH of 4.7–4.9 is desired for most high-ratio cakes. The actual important effect of chlorine appears to be its reaction with the starch to give a modified (oxidized) starch, which can swell to a greater extent than normal (untreated) starch. This gives a batter that is more viscous at the same temperature than the batter of an unoxidized flour. The increased viscosity of the batter keeps the cake from collapsing in the oven and, to some extent, after it is removed from the oven.

Sugars. Sugar increases the temperature at which starch gelatinizes. Thus, it is not surprising that the temperature for the onset of the rapid increase in viscosity increases as the level of sugar in the batter is increased. Perhaps less obvious is the effect of sugar level on batter viscosity. At ambient temperature, viscosity increases as the sugar level is increased. However, at higher temperatures the viscosity is much lower as the sugar level in the batter is increased.

The changes in viscosity can be explained as follows. At ambient temperature, much of the added sugar remains as a solid, as the water is insufficient to dissolve it all. The solid particles are responsible for

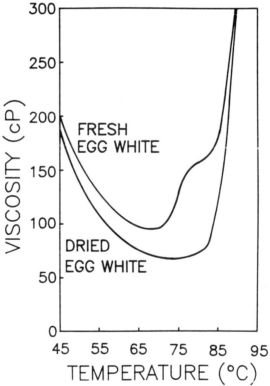

Fig. 16. Viscosity-temperature plots of the minimum viscosity stage of heated AACC cake batters containing fresh and dried egg whites.

the higher viscosity. As the temperature is raised, more of the sugar is dissolved, not only decreasing the number of solid particles but also increasing the amount of solution. Both of these factors decrease the viscosity.

Egg Whites. Egg whites are added to cakes to help build their structure. The proteins in egg white increase viscosity at room temperature by incorporating air. In addition, egg white proteins can set during heating, greatly increasing the viscosity at higher temperatures. As shown in Fig. 16, fresh egg whites are much more effective at doing this than are dried egg whites. It appears that the dried egg whites are denatured during the drying process as they give no denaturation peak in the DSC (Fig. 17).

ANGEL FOOD CAKES

Besides the more standard white layer cakes, we must also consider foam or angel food cakes. A formula for angel food cake is given in Table V. Several things are of interest—for one, the low level of flour. The flour used is weak and is often diluted with wheat starch. The egg whites are the most important component of the angel food formula. Generally, the eggs and sugar are whipped to a protein foam, and the flour is folded in carefully so as not to disturb the foam. The function

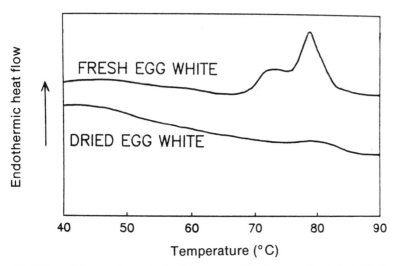

Fig. 17. Differential scanning calorimetric scans of fresh and rehydrated dried egg whites.

of the flour appears to be to provide starch that will gelatinize and thereby remove excess free water. The cream of tartar is added to reduce the pH and thereby improve the whipping of the egg whites. Note that the formula contains no leavening agent. The leavening is produced simply by the air trapped in the egg white foam. Care must be taken not to introduce any fat into the formula, as fat destabilizes the foam. In fact, one cannot use plastic containers because they retain enough fat to destroy the foam.

POUND CAKES

Another type of cake that uses air for leavening is the "pound" cake. The original formula for a pound cake was 1 lb each of flour, butter, eggs, and sugar. This made a very heavy, rich, and expensive cake. A more typical formula is given in Table V. It produces a lighter cake with better eating and keeping qualities.

BEHAVIOR OF BATTER

To produce good light cakes, a batter having many finely divided gas bubbles is formed. As the batter is heated, its viscosity decreases. A good cake batter must retain sufficient viscosity during heating to keep the starch suspended. If the starch, which is dense, settles, a tough rubbery layer forms at the bottom of the cake pan and very light fluffy foam is at the top. The higher viscosity also keeps the air bubbles from colliding with sufficient force to coalesce into larger bubbles. Large bubbles would have sufficient buoyancy to rise to the surface and be lost.

The setting of cakes in the oven appears to be the result of starch gelatinization. As starch gelatinizes, it goes from an inert body that can bind an amount of water weighing about 30% of its weight to one that can bind several times its weight. This increases the viscosity of the batter tremendously, which gives the batter a solid appearance. As discussed in Chapter 12 for bread, transformation of the foam (batter) structure to a sponge structure is necessary to "set" the baked produced. As discussed in Chapter 2, the sugar in the formula controls the temperature at which starch gelatinizes. Consequently, the formula controls at what temperature the cake batter goes from a fluid to a solid. If this occurs at a temperature below the boiling point of water, the cake sets into a solid system. If it is above the boiling point of water, the batter does not set, and the cake collapses into a fudgelike mixture.

Generally, the finer the particle size, the better the cake flour. Therefore, most cake flours are pin-milled to reduce the particle size. Pin-milling also increases the damaged starch in the sample, but this does not appear to effect the cake flour quality.

Biscuits

The American biscuit, actually a chemically leavened bread or bun, is unique to the United States. It has become quite popular, particularly in fast-food establishments, in recent years. The use of chemical leavening gives a dough that results in a rather thick cell wall and a coarse grain in the final product. The flavor is strongly influenced by taste of the soda and the leavening acid. The production process is quite simple: the dough is mixed, sheeted to a desired thickness, cut, and baked.

A variation of the chemically leavened biscuit is the type sold refrigerated and packed in cans. The cans are actually foil-lined cardboard containers. The dough is mixed, sheeted, cut to size, and placed in the can. The volume of dough added to the can is much less that the volume of the can. The ratio of the two volumes must be carefully controlled. After the cans are sealed, they are placed in a proof box at a temperature high enough to trigger the leavening system but low enough not to hurt the dough properties. The dough then "proofs," i.e., expands to fill the can. The excess air that was in the can is expressed through the can, which is permeable to air but not to dough. The dough builds a pressure of about 15 psi and in this condition is quite stable for reasonable periods of time, 60–90 days. Because the dough is not sterile, organisms will eventually destroy the quality of the biscuit. When this starts, the pH drops, the pressure in the can increases to dangerous levels, and the dough becomes very short. Those factors limit the shelf life of the product.

The leavening acid for refrigerated biscuits must meet strict requirements. It must not react during mixing or sheeting of the dough but must react in the proof room. This requirement is met only by a SAPP.

REVIEW QUESTIONS

1. Why are hard wheat flours undesirable for cookies?
2. What are the gases used to leaven baked products?

3. Contrast the use of ammonium bicarbonate and sodium bicarbonate.
4. What are the advantages and disadvantages of ammonium bicarbonate?
5. What are the advantages of sodium bicarbonate?
6. What are the ingredients of a baking powder?
7. What is neutralization value?
8. List the common leavening acids that are used and rank them in order of their reaction temperature.
9. Describe a rotary mold operation.
10. Why are rotary-mold cookies economical to produce?
11. Give an example of cutting-machine cookies.
12. How does the formula for wire-cut cookies differ from that for cutting-machine cookies?
13. What is a typical formula for sugar wafers?
14. What factors determine the quality of a cookie?
15. In general, how is the development of gluten controlled during the mixing of cookie doughs?
16. How much of the sugar in a wire-cut cookie formula dissolves during mixing?
17. Why, when sugar dissolves, is more solution available to make the dough spread?
18. Explain the sequence of events that occurs when cookie dough is placed in the oven.
19. Why do cookies made partly with corn syrup not give a top crack?
20. Explain why cookies become hard and brittle a few days after baking.
21. Outline the process of making saltine crackers.
22. Why is it important to obtain a low pH (4.1) during the sponge fermentation of saltine crackers?
23. What is a desirable pH for baked saltine crackers?
24. What are the typical baking conditions for saltine crackers?

25. How do snack crackers differ from saltine crackers in both formula and processing?
26. How do cakes differ in formulation from cookies? from bread?
27. What are the three major types of cakes, based on the method of preparation?
28. What happens if the viscosity of a cake batter is too low?
29. What is a high-ratio cake?
30. Why is high-ratio cake flour treated with chlorine?
31. How do shortening and surfactants affect the rate of viscosity increase in cake batters during baking?
32. Explain the role of sugar in a cake batter during baking.
33. What is unique about angel food and pound cakes?
34. What controls the temperature at which a cake batter sets?
35. What limits the shelf life of refrigerated dough?

SUGGESTED READING

FARIDI, H. 1988. Flat breads. Pages 457-506 in: Wheat: Chemistry and Technology, 3rd ed. Vol. II. Y. Pomeranz, ed. Am. Assoc. Cereal Chem., St. Paul, MN.

GREENWOOD, C. T., GUINET, R., and SEIBEL, W. 1981. Soft wheat uses in Europe. Pages 209-266 in: Soft Wheat: Production, Breeding, Milling, and Uses. W. T. Yamazaki and C. T. Greenwood, eds. Am. Assoc. Cereal Chem., St. Paul, MN.

HOSENEY, R. C., WADE, P., and FINLEY, J. W. 1988. Soft wheat products. Pages 407-456 in: Wheat: Chemistry and Technology, 3rd ed. Vol. II. Y. Pomeranz, ed. Am. Assoc. Cereal Chem., St. Paul, MN.

LOVING, H. J., and BRENNEIS, L. J. 1981. Soft wheat uses in the United States. Pages 169-207 in: Soft Wheat: Production, Breeding, Milling, and Uses. W. T. Yamazaki and C. T. Greenwood, eds. Am. Assoc. Cereal Chem., St. Paul, MN.

MANLEY, D. 1991. Technology of Biscuits, Crackers and Cookies, 2nd ed. Ellis Horwood, New York.

MATZ, S. A., and MATZ, T. D. 1978. Cookie and Cracker Technology. Avi Publishing Co., Westport, CT.

NAGAO, S. 1981. Soft wheat uses in the Orient. Pages 267-304 in: Soft Wheat: Production, Breeding, Milling, and Uses. W. T. Yamazaki and C. T. Greenwood, eds. Am. Assoc. Cereal Chem., St. Paul, MN.

WADE, P. 1988. Biscuits, Cookies and Crackers. Vol. I. Elsevier Applied Science, London.

Glass Transition and Its Role In Cereals

CHAPTER 14

For many years, when cereal chemists discussed polymers, they were speaking of synthetic plastics. Only in recent years have we begun to understand that proteins and polysaccharides are also polymers. As chemists, we have always known that these molecules were polymers, but somehow we felt that they were different from the synthetic ones. Therefore, few were concerned with the polymer literature and what could be learned from it.

It is certainly true that proteins, which are polymers of some 20 different amino acids, are different from polyethylene, which is made up of a single monomer unit. However, the chemical and thermodynamic rules that apply to the polyethylene apply equally to proteins.

In trying to understand the behavior of a polymer, one of the most important properties to know is its *glass transition temperature* (T_g). This value affects many of the physical properties of the material. A low T_g means that at room temperature the polymer is rubbery and flexible and at higher temperatures may even flow under the force of gravity. On the other hand, a glass transition that occurs at higher temperature means that the polymer material is rigid and perhaps fragile or brittle at room temperature.

Much can be learned from the generalized plot in Fig. 1. It relates the rheological properties of a polymer to its temperature. As the temperature is raised through the glass transition, the modulus drops. The physical properties of the polymer go from glassy to leathery. Increasing the temperature causes the material to become rubbery. This temperature region is called the rubbery "plateau" and, as the name implies, it extends over a range of temperature. The length of the rubbery plateau varies depending upon several factors, the most important of which is whether the polymer is cross-linked or not. As the temperature is raised, noncross-linked polymers flow, but cross-linked polymers remain rubbery. Increasing the temperature even higher causes the polymer to thermally degrade.

What is a Glass Transition?

Stated in very simple terms, a glass transition is a large change in modulus that a material undergoes at a particular temperature (or, more accurately, over a narrow temperature range). The classic example is glass. When the glass blower heats a piece of rigid, brittle glass, it softens and eventually flows. If the flexible material is allowed to cool, it again becomes rigid and brittle. The cooling phase is called *vitrification*. The change in modulus that characterizes the glass transition is amazingly constant for most pure polymers at about five orders of magnitude over a relatively narrow temperature range (Fig. 1).

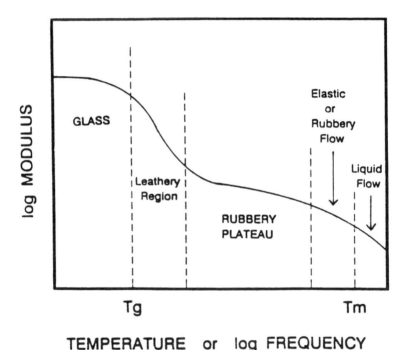

Fig. 1. Master curve of the modulus as a function of temperature or frequency, illustrating the glass transition temperature (T_g) and the five regions of viscoelastic behavior of a partially crystalline polymer. An amorphous polymer would be similar except that the melting temperature (T_m) would be missing.

The midpoint of the modulus change is often taken as the glass transition temperature and designated T_g. However, the glass transition is a *secondary transition*. This means that the T_g is not a truly reproducible temperature like a melting point (a *primary transition*). The glass transition temperature is affected by the history of the sample (how long it has been stored, rate of cooling, etc.).

Glass transition is a property of *amorphous* (noncrystalline) materials. Thus, a crystalline material does not show a T_g. Many polymers are partially crystalline, and only the amorphous part of these polymer shows the glass transition.

What Causes Glass Transitions?

Picture a polymer system consisting of relatively high-molecular-weight polymers that are highly entangled. A ball of worms or snakes may bring the picture to mind. At moderate temperatures, each polymer chain undergoes a number of motions. Various side chains are spinning, etc. In polymers, an important type of molecular motion is the segmental motion of the chain backbone. This is a long-range motion that involves a cooperative motion along the backbone chain. It can be visualized as a snake crawling. This motion of the chain defines the volume of the polymer. The polymer chain requires this amount of space and, by its motion, keeps other chains from invading its space.

As the temperature of the system is lowered, the motion of the chains is slowed. As the motion slows, the individual polymer chains are less able to keep other chains from invading their space. The result is a net reduction in the volume of the system and an interaction of the polymer segments with each other. As the polymer segments of the various polymers interact with each other, the complex viscosity of the system increases rapidly. At some point, the viscosity increase becomes large enough that the chains no longer have segmental motion. Now the kinetic energy of the system is no longer enough to overcome the friction of the chains interfering with each other. The system is now vitrified (glassy). As the sample is heated, the process is reversed: the kinetic energy of the chains increases to a point where it overcomes the friction of their rubbing on each other; the volume increases; and the physical properties go from glassy to leathery, to rubbery, and finally to a viscous flowable material. All of this occurs over a relatively narrow temperature range.

The above is an explanation for the glassy transition phenomenon based on *free volume theory*. There are other explanations, but they are beyond the scope of this book. Note that this explanation requires the volume of the system to change as it goes through the glass transition. Also note that the vitrification process (making the material glassy) does not require the stopping of other motion in the polymer system. For example, the spinning of the side chain continues even though the material is glassy.

Factors That Affect the Glass Transition Temperature

A number of factors can affect the T_g of a polymer. Most of the them are intuitively obvious if we remember the free volume theory that explains the glass transition and how the various factors affect the free volume of the polymer.

The number, size, and rigidity of substituents (i.e., side chains) on the polymer chain affect the glass transition. An increase in the number or size of the substituents (if they are rigid) increases the T_g. If they are not rigid, these side chains may act as a plasticizer and thereby actually lower the T_g.

Pressure is another factor known to affect the T_g. High pressure decreases the T_g because high pressure decreases the free volume. However, to have a significant effect, the pressures must be large (on the order of hundreds or thousands of atmospheres).

The molecular weight of the polymer can also affect its T_g. This is explained by two factors. First, the ends of a polymer are less restricted and thus can increase the free volume. Second, as the molecule becomes larger, the ends of the polymer are less of the total. Generally, the T_g decreases from a value for a polymer of infinite molecular weight by a function of the inverse of the molecular weight. Cross-linking of polymers is another way of increasing the molecular weight. That would be expected to increase the T_g, which it has been shown to do. In addition, the cross-links also interfere with the movement of the polymer chain, which also increases the T_g. Partial crystallization of the polymer can act as a form of cross-linking and increase the T_g. Although this is generally true, it has been shown to not always be the case, as some crystallized polymers show a lower T_g than do their uncrystallized counterparts.

Measurement of Glass Transitions

As mentioned above, the volume of the polymer material changes as glass transition takes place. When an amorphous or partially amorphous material is heated, the volume of the polymer increases with temperature in a linear fashion until the T_g is approached. A that point, the volume increases at a much faster rate. Once the transition has occurred, the slower linear increase resumes. This is in contrast to a crystalline material, for which the increase in volume occurs sharply as the melting point is reached. The change in volume for amorphous and partially crystalline materials is shown in Fig. 2.

This increase in volume can be used to determine T_g. An instrument such as a thermal mechanical analyzer measures the change in dimension as the material is heated. One problem with this type of instrument is that it measures only the change in dimension and does

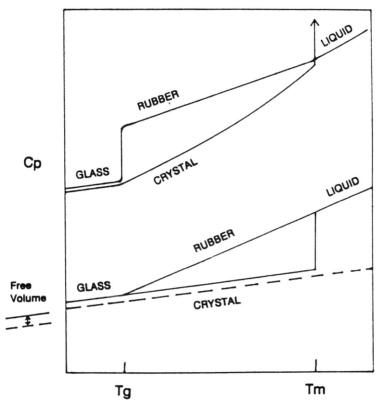

Fig. 2. Plot of heat capacity (C_p), top curves, and free volume, bottom curves, as a function of temperature for various polymers. An amorphous glassy polymer goes from glassy to rubbery to liquid. For this polymer, the C_p changes rapidly at the glass transition temperature (T_g) and then increases gradually with increased temperature. The free volume increases gradually at temperatures above T_g. For a crystalline polymer, there is a step change in both C_p and free volume when the crystal melts (T_m). For a partially crytalline (and therefore partially amorphous) polymer, both of the above changes in C_p and free volume are seen.

not tell us what is responsible for that change. If it is due to the glass transition, we have no problem. However, if gas trapped in the system is expanding or if solvent in the polymer is vaporizing, the resulting volume change may be mistaken for glass transition.

Another popular method of measuring glass transitions is with differential scanning calorimetry (DSC). As a polymer becomes mobile by going through a glass transition, its heat capacity increases. This change in heat capacity can often be measured with the differential scanning calorimeter (Fig. 2). Because the glass transition is reversible, the change in heat capacity can be measured in either direction (heating or cooling) with DSC.

DSC works relatively well with pure materials. However, if other materials are diluting the polymer you want to measure, the small change in heat capacity may be too small for the sensitivity of the calorimeter.

A number of rheological and electrical methods can measure glass transitions. In general, they have not been used with cereal systems and thus are outside the scope of this book.

Effect of Plasticizer (Water) on Glass Transitions

The T_g of a polymer is also affected by the level of solvent or *plasticizer* in the polymer. The amount of friction between the polymer chains is controlled by the amount of plasticizer in the system. Therefore, if a polymer is glassy at room temperature and a plasticizer is introduced into the polymer matrix, the polymer goes through the glass transition with no change in temperature. It has been plasticized. In effect, the T_g is lowered to room temperature or below.

While the polymer is glassy, the rate of diffusion of small molecules through the polymer is slow. Thus, adding a plasticizer to the system takes time. Not all material added to the system will act as a plasticizer. If the friction between the polymer chains is increased rather than decreased, it follows that the added material acts as an antiplasticizer and, of course, increases the T_g.

Glass Transitions in Cereals

The major polymers in cereal grains are the proteins, starch, and small amounts of pentosans in the endosperm and the pentosans and cellulose in the pericarp. The proteins and pentosans are amorphous polymers, whereas the starch and cellulose are partially crystalline. Glass transitions are undoubtedly important to the properties and behavior of all these polymers.

Examination of a mature wheat kernel at normal safe-storage moisture contents (<14%) shows the pericarp to be fragile and the endosperm to be hard and brittle. If the kernel is soaked in water (a plasticizer), amazing changes result. The pericarp becomes leathery and tough. The endosperm also changes by becoming soft and rubbery. These are classic examples of behaviors that result from going through a glass transition. In the above cases, the polymers in the kernel have made the transition with no change in temperature.

The continuous phase within the cells of the wheat kernel consists of amorphous gluten proteins. Because of this, the changes noted above in the wheat endosperm are likely the result of changes in the gluten proteins. If we examine dry gluten, particularly a sample that is not porous, we find that it is hard, hornlike and, definitely glassy. The gluten proteins have been shown to go through the classic changes associated with glass transition (see Fig. 1, Chapter 10). It has also been shown that small amounts of water greatly decrease their T_g (see Fig. 8, Chapter 10). At about 16% moisture, gluten proteins go through the glass transition at room temperature.

As often stated, the gluten proteins are unique in many ways, especially in their ability to form a viscoelastic dough when hydrated.

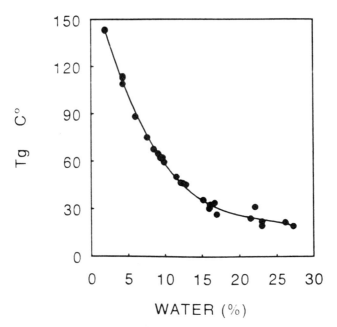

Fig. 3. Glass transition temperature (T_g) of zein as a function of moisture.

Much has been written about the reasons for this unique ability, including the role of specific amino acids in the proteins, the proteins' three dimensional structures, and many others. However, from a polymer science viewpoint, those items are essentially beside the point. To have a rubbery material, all that is needed is a large polymer that has a T_g at the correct temperature. In the case of gluten, the T_g is at room temperature at 16% water. Addition of more water moves the polymer into the leathery and finally into the rubbery region.

From the above facts, we must conclude that the uniqueness of wheat gluten is that it undergoes a glass transition at the "correct" temperature. In addition, gluten has the ability to absorb large amounts of water, as it is a water-compatible polymer.

In contrast to wheat endosperm, corn endosperm does not change appreciably when it is soaked in water. It hydrates but does not swell appreciably and retains its hard, gritty, nature. This phenomena can be explained by the fact that the matrix protein in corn is highly cross-linked and thus does not absorb large amounts of water. A second factor is that the glass transition in corn protein (Fig. 3) occurs at a higher temperature. Recently it was shown that mixing zein and starch at elevated temperatures (above zein's T_g) gives a cohesive dough in a farinograph (Fig. 4).

STARCH

Starch, as it occurs in nature, is a partially crystalline polymer system (see Chapter 2). When starch is gelatinized, all the crystals are melted

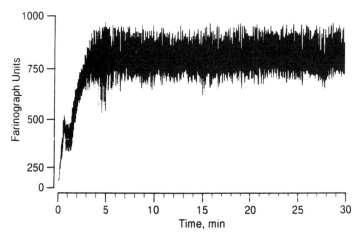

Fig. 4. Farinogram of a zein-starch dough mixed at 35°C.

and it freely hydrates and swells in water. Under these conditions, starch gives a classic glass transition curve, showing a rapid decrease in T_g as the water content is increased. With excess water, the glass transition has been shown to be about $-7°C$.

The story with native starch is complicated by two factors. First, the starch occurs in granules, and these limit the uptake of water. Starch takes up only about 30% water at room temperature. The second factor is the crystalline nature of the starch. Generally, cross-linking (crystals are a special type of cross-link) raises the T_g. In pregelatinized starch, the cross-links are no longer there, and the T_g shifts to a lower temperature. This is shown in Fig. 5. However, this figure and what the data in the figure suggest have not been universally accepted. Other workers have suggested that the glass transition of native starch occurs at higher temperatures. A thorough discussion of the arguments appears to be beyond the scope of this book. Likely it will be some time before a consensus is reached on the glass transitions in native starch.

OTHER POLYMERS IN CEREALS

There have been no literature reports concerning glass transitions in cellulose or pentosans in cereals. However, it does appear obvious that they exist and may be quite important in the processing of grains.

Importance of Glass Transitions in Cereal Products

A loaf of bread comes from the oven with a hard, crisp crust that is fragile and, in the author's opinion, truly delightful. However, if the loaf is placed in a plastic bag after it is cooled, the crust becomes tough and leathery within a very short time. The crust has gone from delightful to almost inedible within a short time. What is responsible for the change?

Coming from the oven, the crust is very dry; it has perhaps 2% moisture. With time, the moisture from the crumb migrates to the crust, as an equilibrium is established. At the same time, moisture is lost to the atmosphere, so the crust's property is preserved. However, if the bread is placed in a plastic bag, as all American bread must be (although the reason is beyond me) the loss of water from the crust is slowed. As a result, the moisture content of the crust increases.

Increasing its moisture content makes the polymers in the crust go from glassy to leathery even though the temperature stays constant. Toasting the bread drives the moisture out and restores the glassy state to the crust and, indeed, to the rest of the bread if sufficient moisture is removed.

There are many other examples of glass transitions in cereal products: corn chips (a nice glassy product) that lose much of their desirability if the moisture content increases and the product goes from glassy to leathery; breakfast cereals that become soggy in milk before they are consumed; pizza crust that becomes soggy; popcorn left to pick up moisture becoming quite tough and not nearly as desirable. A saltine cracker is another product that gets much of its characteristic delightful texture from the fact that it is glassy at room temperature, with a moisture content below about 2%. Breading and batter products that are used to coat fish, chicken, or other products are also crisp when dry and soggy at higher moisture.

Another good example of the occurrence of glass transition in cereal products is the use of a cooker extruder, for instance, to make extruded snacks. Corn or other cereals in the glassy state are fed into the extruder, and the moisture and temperature are both raised to cause the cereal material not only to go through a glass transition but also to become a flowable mass. As the material comes from the extruder, some of the moisture is lost and the plastic material cools to again form a glassy product.

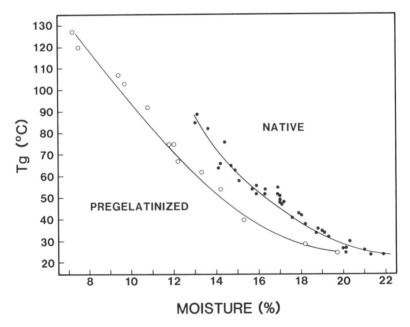

Fig. 5. Glass transition temperature (T_g) of native and pregelatinized wheat starches as a function of moisture.

The properties of cereals and many other materials can be summarized in the generalized "state diagram" given in Fig. 6. From this plot we see that if a polymer product starts at point A in the glass region and we raise the temperature of the product, it will first become leathery, then rubbery, and eventually will flow in most cases. Likewise, if the temperature remains constant and we raise the moisture content, the product can also go from glassy to leathery and rubbery. Such a diagram is very useful in understanding the properties of cereal products.

Glass Transitions of Sugar Solutions

Although we usually think of polymers when we think of glass transitions, it is also true that smaller molecules can undergo glass transitions. Sugar solutions that are used with cereal products are a good example of these. The sugar coating that is used for some of the ready-to-eat breakfast cereals is a sugar glass. The glassy material makes diffusion

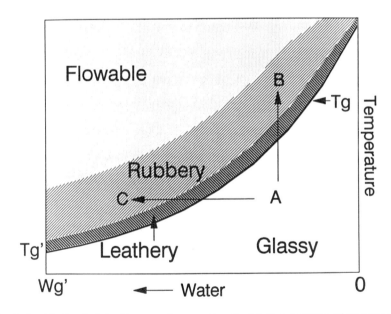

Fig. 6. An idealized-state diagram of a cereal material as a function of temperature and moisture. T_g is the glass transition at no moisture and $T_{g'}$ is the glass transition at the moisture content at which free water (freezable) starts to appear as a separate phase ($W_{g'}$). A, assumed starting point of polymer; arrow to B, effect of increased temperature; arrow to C, effect of increased moisture.

of water very slow and thus makes the cereal product last much longer under adverse conditions. A similar sugar glass is used to protect popcorn in popcorn balls or in the old favorite snack, Cracker Jacks.

REVIEW QUESTIONS

1. Explain how the modulus of a glassy polymer changes as the temperature is raised.
2. Explain how the modulus versus temperature curve changes for cross-linked polymers as opposed to noncross-linked polymers.
3. What is meant by the statement that glass transitions are secondary transitions?
4. In a polymer that is partially crystalline and partially amorphous, which part is responsible for the glass transition?
5. What is vitrification?
6. Explain the free volume theory and how it applies to glass transitions of polymers.
7. Explain how and why the molecular weight affects the glass transition temperature.
8. Is it true that all chemical motion is greatly slowed in a glassy material?
9. What are two popular methods of measuring glass transitions?
10. How does water act as a plasticizer?
11. Explain how a polymer can go through a glass transition at room temperature.
12. Explain from a polymer science viewpoint why wetting wheat (tempering or conditioning) before milling is important to the miller.
13. Explain from a polymer science viewpoint why wetting flour is essential to forming a dough.
14. From a polymer science viewpoint, explain why wheat is unique in forming a viscoelastic dough.
15. Explain why a mixture of corn flour and water does not form a viscoelastic dough.

16. Give three examples of the importance of glass transitions in cereal products.

17. Explain how a "state diagram" can help explain the physical properties of cereal products.

18. Explain how popcorn is kept crisp in Cracker Jacks.

SUGGESTED READING

EISENBERG, A. 1984. The glassy state and the glass transition. Pages 55-96 in: Physical Properties of Polymers. J. E. Mark, A. Eisenberg, W. W. Graessley, L. Mandelkern, and J. L. Koenig, eds. American Chemical Society, Washington, DC.

LEVINE, H., and SLADE, L. 1992. Glass transitions in foods. Pages 83-222 in: Physical Chemistry of Foods. H. G. Schwartzberg and R. W. Hartel, eds. Marcel Dekker, New York.

VICKERS, Z. M. 1988. Crispness of cereals. Pages 1-19 in: Advances in Cereal Science and Technology, Vol. 9. Y. Pomeranz, ed. Am. Assoc. Cereal Chem., St Paul, MN.

Pasta and Noodles

CHAPTER 15

Pasta and noodles are wheat-based products that are formed from a dough but are not leavened. The processes by which they are formed are quite different, as are the types of flours used. The formulation is generally very simple, often only flour and water for pasta; flour, water, and salt for oriental noodles; and flour, water, salt, and egg for U.S. noodles.

Pasta

Pasta, paste, and *alimentary pastes* are terms that describe a large number of products (Fig. 1). For this discussion, the terms will be restricted to the extruded products that are generally made from durum wheats. This effectively eliminates the various noodle products. Even with such a restriction, the number of shapes and sizes made are astounding. Some manufacturers regularly make 60–70 different products. The most common types are macaroni (U.S. standards state that it must be tube-shaped, hollow, and more than 0.11 in. but not more than 0.27 in. in diameter) and spaghetti (cord-shaped, not tubular, and more than 0.06 in. but less than 0.11 in. in diameter, according to U.S. standards). The consumption of pasta is nearly worldwide but the annual consumption varies quite widely. U.S. consumption is about 10 lb per person per year, whereas the consumption in Italy is five to 10 times that amount.

It is generally believed that the ideal raw material for pasta is durum semolina. Before we can discuss an ideal raw material, we must define an ideal product. The uncooked pasta must be mechanically strong so that it will retain its size and shape during packaging and shipment. It should also be uniformly yellow, as consumer acceptance has been strongly linked to a uniform, translucent, yellow color. When cooked in boiling water, the product must maintain its shape with no splitting or falling apart. Also, after it is cooked, the pasta should give a firm bite (the so-called "al dente") and the surface should not be sticky. The cooking water should be free of starch. Finally, the pasta should be resistant to overcooking.

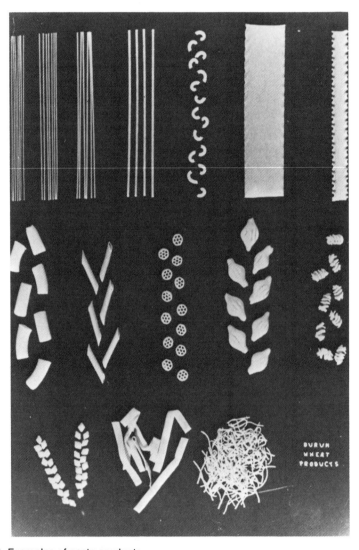

Fig. 1. Examples of pasta products.

Durum wheats are generally believed to come closer to producing such a product than other types of wheats. Several factors appear to support the belief. In years when durum wheats are in short supply because of disease or other factors, the price of durum wheat is high. That leads to more use of common wheats in pasta products. Generally, the consumption of pasta then declines. As the supply of durum again increases, pasta consumption also increases.

The durum wheats differ from the common wheats. They are tetraploids and the common wheats are hexaploids. Durums are mostly spring wheats, although winter durums are known. They are usually amber in color; actually, they are white wheats with a translucent endosperm that gives them an amber appearance. Red durum wheats are also known, but they are used for feed and not for semolina production. Endosperms of durum wheats are high in carotenoid pigments (Fig. 2); these are what give the pasta its yellow color. Because of the relationship between the yellow color and consumer acceptance, the amount of pigmentation has been used as a selection tool for good-quality durum. It may well be that the durum wheats originally contained no more pigments than did the other types of wheats but now have more because we have selected for the pigments.

In general, durum wheats do not make bread. Durum gluten is usually weaker than common wheat gluten. However, the more recently developed cultivars have stronger gluten types and thus produce a better loaf of bread, although it is still poor by common wheat standards. Interestingly, the durums with stronger gluten also generally give pasta with a stronger "al dente." Occasionally, in certain parts of the world (for example, North Africa and India), one can find durum wheat being used to produce bread.

Probably the most outstanding characteristic of durum wheats is their hardness. The grain is physically very hard, much harder than the hard common wheats (Fig. 3). The wheat can be milled to give good yields of *semolina*, which is the purified middling from durum wheat. Durum wheat is so hard that it is difficult to reduce to a flour

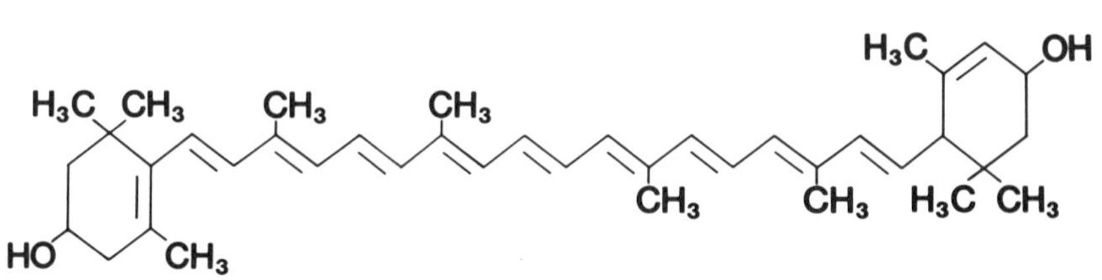

Fig. 2. Structure of a carotenoid pigment (α-dihydroxy carotene).

fineness. When it is reduced to flour, the percentage of damaged starch is several times higher than that found with common wheats. The objective of the durum milling process is to produce as much semolina as possible. The process uses a large number of purifiers. Flour is also produced, but in general, it is of lower value than the semolina. Durum flour is often used to make noodles, but it can also be used to make pasta; it generally gives excellent products except that they are not as resistant to overcooking as are products made from semolina.

In North Africa, durum semolina is used to make the product *couscous*. The semolina is steamed and agglomerated into particles of about 2 mm in diameter. The product may be dried and resteamed before being eaten or may be eaten directly. *Couscous* is usually eaten with a sauce.

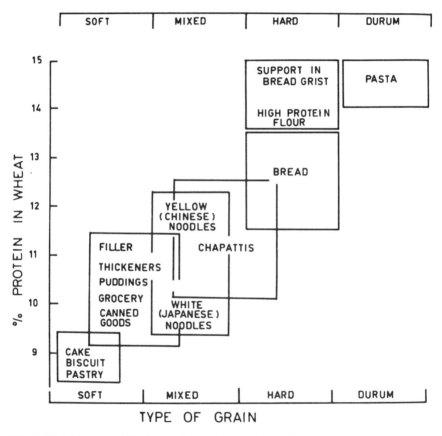

Fig. 3. Wheat types and the types of products made from them.

When durum wheat is expensive, common hard wheat *farina* (purified middlings from hard wheat) is often blended with durum or used by itself to produce pasta. In general, the hard-wheat farina produces good pasta except that it does not have the yellow color and is not as resistant to overcooking as pasta produced from semolina.

The Production Process

In outline form, the process is simple. Water is added to semolina to obtain about 31% moisture. The mixture is kneaded to obtain a homogeneous mass and then extruded through a die, dried, and packaged. In actuality, both the mixing and the drying are rather complicated.

At 30% moisture, the dough produced is very dry. That level of water is less than one half of the level used in a bread dough. When mixed, the dough forms into balls of about 1 in. (2.54 cm) in diameter. The size of the balls is diagnostic for the correct amount of water. Too much water gives larger balls and eventually a continuous dough, whereas too little water produces smaller balls.

Mixing is performed in an airtight mixer in the absence of air. Air in the mixer is detrimental for two reasons. First, as the dough is forced down the barrel of the extruder, air is dissolved in the aqueous phase of the dough. When the pressure is removed as the dough exits the die, small bubbles can appear in the extruded piece. These small air bubbles make the piece appear opaque rather than translucent, which interferes with the perception of the yellow color. Also, the air bubbles cause a point of weakness in the dried product.

The second problem associated with the presence of air concerns the enzyme lipoxygenase. All flours contain some lipoxygenase activity. In general, the durum wheats have been selected for low levels of the enzyme and contain much lower levels than are found in the common wheats. This is a major reason why the hard wheat farina will not give yellow pasta—the enzyme bleaches the yellow carotenoid pigments. To do this, lipoxygenase requires polyunsaturated free fatty acids and oxygen. The grain almost invariably contains the free fatty acids; thus, we attempt to control the bleaching action by keeping the oxygen content as low as possible.

From the mixer, the dough enters an auger that kneads and exerts pressure on it as it moves down the barrel of the extruder to the die. The combined effects of the kneading and pressure produce a smooth, homogeneous dough that can be extruded. A considerable amount of unwanted heat is produced in that process; therefore, the extruder barrel is jacketed and cooled with water. In general, the temperature of the dough is maintained at less than 45°C. Because both the temperature and the moisture content of the dough are low, essentially no expansion is obtained as the product exits the die.

The dies are normally made of bronze, although both stainless steel and Teflon-coated dies are also used. Bronze dies give excellent products but tend to wear out rapidly. The product is abrasive and wears out the soft bronze, thus giving misshapen products. Bronze dies must also be cleaned thoroughly, or frozen when not in use. If they are not, the bacteria in the dough produce acids that pit the die, also leading to inferior products. Both stainless steel and Teflon dies are smoother than bronze; therefore, the production rate is faster and the product is smoother and appears more yellow. However, the change in the surface character also changes the cooking characteristics. The product cooks more slowly; water penetration is slower; and the surface tends to become mushy.

The extruded product still contains about 30% moisture and must be dried to about 12% before it is stable for shipment and storage. If drying is too fast, the product may check. *Checking* is the formation of numerous hairline cracks in the product that make it appear opaque and also decrease its strength. The checking is caused by differential contraction as moisture is removed from the product. On the other hand, drying too slowly can also cause problems. Long goods (such as spaghetti) stretch under their own weight. If not dried, all products turn sour or mold given sufficient time.

The standard drying procedure is to rapidly dry the outer surface of the piece. This gives strength and decreases the chances of mold growth. Usually about 40% of the total water in the piece is removed over a 30-min period, giving a relatively dry area on the outside while the interior remains moist. This is sometimes referred to as *case hardening*. Following that rapid drying is a sweating period, in which the product is held in relatively humid air and the moisture content is allowed to become more uniform. The time is usually 2-4 hr at 90% rh. The sweating period is followed by a final slow drying to about 12%. The final drying may take 10-16 hr. The use of microwave energy to dry pasta appears to be gaining in popularity. Microwaves have the advantage of heating the water in the piece uniformly; thus, the wet and dry gradients that cause checking are not produced. The time required today is much less with the use of microwaves.

For shipment and storage, pasta must be dried. However, eating fresh pasta (cooked without drying) will convince one that drying is not in itself a desirable step.

Noodles

Noodles are a type of pasta that is generally made from flour, rather than semolina or farina, and contains salt in addition to flour and water. The U.S. standard of identity for noodles states that they must be made of wheat dough containing eggs. Dried noodles must contain less than 13% moisture and more than 5.5% egg solids.

Noodles are thought to have originated in China and are still a popular food throughout Asia. As much as 40% of the wheat consumption in Asia is as noodles. In general, oriental noodles contain no eggs.

There are many types of oriental noodles (Figs. 4 and 5). They vary in their ingredients, degree of cooking before sale, and/or degree of drying. Fresh noodles are not cooked before sale and contain about 35% moisture. Thus, they deteriorate rapidly unless stored under refrigeration. After 50-60 hr at refrigeration temperatures, the noodles darken and become moldy. For good consumer acceptance, they must be white or light yellow. The darkening with storage is thought to be caused by polyphenoloxidase, the enzyme that causes enzymatic browning of many cut fruits. Fresh noodles are usually produced with

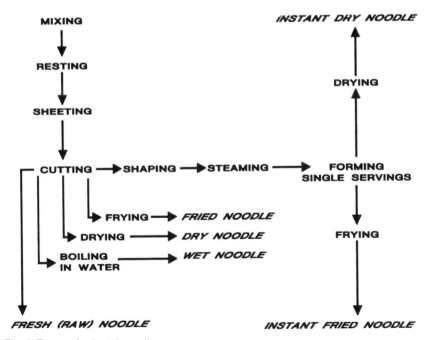

Fig. 4. Types of oriental noodles.

a thin cross section, which allows them to cook rapidly. They are made from relatively strong flours, so they can be handled in the wet form. This gives the cooked noodles a chewy texture.

Wet noodles are cooked in boiling water before they are sold. Once cooked, they have a moisture content of about 52% and thus have a relatively short shelf life (40 hr at room temperature). The boiling denatures the polyphenoloxidase enzyme; therefore they do not turn brown during storage.

Chinese wet noodles are made with a relatively weak (all-purpose) flour and *Kan-sui*. *Kan-sui* is composed of sodium and potassium carbonates. It replaces sodium chloride in the formula. The carbonates produce an alkaline dough that gives a strong noodle with a bright yellow color. The color comes from the flavonoid pigments.

Dry noodles are not precooked but are formed at a moisture content of about 35% and then dried to 8–10% moisture. The drying is traditionally done in the sun but can also be done under controlled low humidity. Because the noodles are dry, they can be easily handled and have a long shelf life. In dry noodles, color and opaqueness are the

Fig. 5. Oriental noodles. Top row: various dried noodles; bottom row: left, wet noodles; center two, ramen; right two, packaged wet noodles.

major quality factors. Dull, gray, or brown noodles are considered inferior. In addition, the noodles should have a uniform shape and cleanly cut sides. Because of the importance of color, highly purified flours are generally used.

Instant fried noodles are cut, waved, precooked with steam, formed into individual servings, and dried by deep-fat frying to 5–8% moisture. They are called *ramen* or *ramyeon* and are usually packaged with seasonings. The instant fried noodles have a taste distinctive from that of the other noodles, probably because they pick up about 20% fat during frying.

For good consumer acceptance, instant fried or ramen noodles should be free of rancidity. Consumers also prefer a white color, which requires the noodles to be fried for a short time, at a relatively low temperature, in a good-quality oil. When ramen noodles are cooked by the consumer in boiling water, no fat should separate in the cooking water. After cooking, the noodles should have a relatively strong bite and a firm, nonsticky surface. Because fried noodles are relatively dry, they have good storage properties, particularly if packaged to exclude oxygen and light.

Steamed and dried noodles are precooked with low-pressure steam, shaped, and then dried to about 10% moisture. Drying is generally in the sun or with a convection oven. Because they are dry, they have a good shelf life (approximately one year). This type of noodle requires a longer cooking time than do dried noodles of a similar size. The taste and texture are different from those of the fried or dry noodles.

As with all foods, wide variations exist in the type of noodle preferred by different peoples. In Korea and China, noodles with a chewy texture are preferred. In Japan, a softer texture is desired. Thus, the Japanese-style fresh and wet noodle (*udon*) is popular there. In all Asian noodles a smooth surface is highly desirable.

Starch noodles, made principally from mung bean starch, are popular throughout Asian countries. The noodles are made by allowing a starch-water slurry to flow through a small orifice into boiling water. This results in vermicelli-sized strings. The noodles are then allowed to retrograde to gain strength. The consumption of starch noodles is small compared to the consumption of flour noodles.

Flour for Noodles

Because flour makes up 95–98% of the dry solids of noodles, its importance appears obvious. Although flour consists of many components, this discussion concentrates on four of them (pigments, protein, starch, and enzymes) and their importance in noodles.

The two major types of pigments in wheat flour are carotenoids and flavonoids. *Carotenoid pigments* are found in the endosperm and give flour its creamy-yellow color. The amount of carotenoids varies widely among different cultivars of wheat. These pigments can be bleached rather easily and are destroyed by bleaching agents added to flour. Two such agents that are commonly used are benzoyl peroxide and enzyme-active soy flour (in which the active ingredient is the enzyme lipoxygenase). On the other hand, *flavonoid pigments*, which come mostly from bran contamination of flour, are not bleached by the normal bleaching agents. The flavonoids (Fig. 6) are relatively stable and are colorless at acidic pH but give a yellow color at high pH. They are the source of the yellow color in Chinese noodles containing *Kansui*. The presence of iron salts with the flavonoids at alkaline pH often gives rise to green and brown colors. Because of contamination by the flavonoid pigments in low-grade flours and the resulting problems, noodles are often made from low-extraction flours.

The unique ability of wheat flour to form a cohesive, elastic, and extensible dough is the result of gluten proteins in the wheat flour. Protein quantity and quality are important in noodle making. High amounts (10–14%) of strong protein produce noodles with a chewy, elastic texture. Flour with too low a protein content gives noodles with poor cooking tolerance. When overcooked, the noodles are mushy and sticky.

Starch is the predominant component in flour. It changes from the raw granular form to the gelatinized form, which causes the noodle to set. If the starch on the surface of the noodle is overcooked before the interior becomes cooked, the surface of the noodle becomes sticky. When a noodle is properly cooked, the starch on the surface is not sticky but gives a smooth mouth-feel.

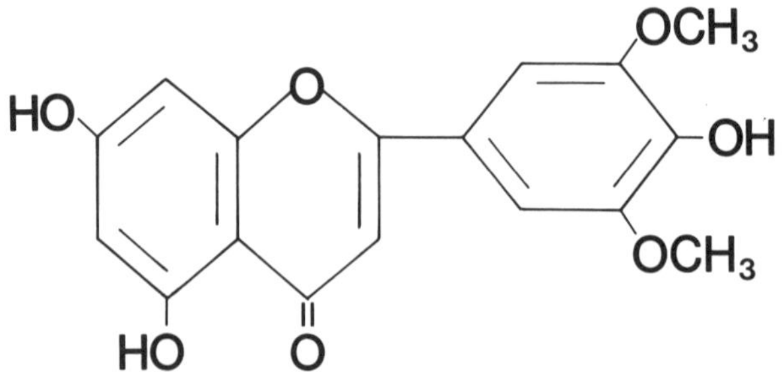

Fig. 6. Structure of a flavonoid pigment (tricin) from wheat flour.

Although flour contains only small amounts of enzymes, those present can significantly affect noodle quality. Excessive amounts of α-amylase (caused either by sprout damage or malt supplementation) in flour cause rapid breakdown of the noodle structure. Polyphenoloxidase is another important enzyme affecting noodle quality, as discussed earlier.

Noodle Making

As with other pasta procedures, noodle making is relatively simple. The dry flour is placed in the mixer, and the water and salt are added. The amount of water is limited, usually less than 35%, based on the flour weight. That level of water is not enough to form a dough at first. Instead, the flour forms balls or spheres. The mixer is designed to cut dry flour into the water-rich spheres. Mixing generally takes 5–10 min and has as a primary objective the uniform distribution of water.

After mixing, the dough is allowed to rest for 10–15 min, which also helps to distribute water evenly throughout the flour particles. The crumbly dough is then pressed between two large-diameter rolls to produce a dough sheet about 1 cm thick. The dough sheet is rolled thinner by seven to 10 successive passes through reduction rolls. After reaching the desired thickness of 1–2 mm, it is passed through a pair of cutting rolls (Fig. 7).

The function of sheeting is to form a sheet of dough with a uniform thickness and gluten development. Sheeting operations develop bread doughs in much the same way that mixers do. Because the noodle dough is always sheeted in the same direction, the gluten fibrils are aligned in the direction of sheeting. That alignment gives the noodles more strength in their long rather than their short direction.

Two important factors in noodle-dough sheeting are the sheeting speed and the sheeting ratio. The *sheeting speed* is the speed of the rolls, or how fast the dough passes through the rolls. The *sheeting ratio* is the thickness of the dough after sheeting divided by the thickness of the dough before sheeting. Both of these variables must be controlled to obtain a good noodle.

Noodle cutting is performed by a pair of slotted rolls aligned so that they run point to point, thereby producing a noodle with a square cross section (Fig. 7).

Dry noodles are produced by hanging the freshly cut noodles (2–3 m long) over wooden dowels. The loaded dowels are then placed in an enclosure for drying. The goal is to dry the noodles as soon as possible and to produce a white, strong, uniformly shaped noodle.

Because noodles contain salt, moisture loss is not as fast as for spaghetti and other pasta, and therefore checking is not as serious a problem with noodles. However, noodle drying is still done in the same three-stage regime as is used for pasta.

Steamed noodles are produced to give instant products. Steaming is preferred to cooking in boiling water because no solids are lost in the cooking water and the noodles retain their shape. During steaming, if two noodles touch, they are bonded together. Therefore, with proper waving and shaping, individual servings can be produced that are continuous but with an open network.

Fig. 7. Noodle machine sheeting and cutting noodles.

Frying is one of the fastest ways to remove moisture from noodles and thus fix their structural form. During frying, water is removed rapidly by the hot oil. That results in a porous noodle structure that rehydrates quickly when water is added. That, of course, is a important property of instant noodles. Frying is usually at 140–150°C for 0.5–2 min.

REVIEW QUESTIONS

1. What is the composition of most pasta products?
2. How is macaroni defined?
3. How is spaghetti defined?
4. What is semolina?
5. What are the qualities of a good pasta product?
6. How do durum wheats differ from common wheats?
7. How does one determine the level of water to add to a pasta dough?
8. Why is mixing done in a vacuum?
9. How does the enzyme lipoxygenase bleach flour?
10. What are the advantages and disadvantages of bronze dies in a pasta extruder?
11. How is checking controlled during the drying of pasta?
12. What is the composition of U.S. noodles according to the standards of identity?
13. What is the composition of oriental noodles?
14. What is *Kan-sui* and where is it used?
15. List the various types of oriental noodles.
16. What pigments are responsible for the yellow color of noodles?
17. What makes the surface of a noodle become sticky during overcooking?
18. How are noodles made commercially?

19. Explain how sheeting speed and sheeting ratio are important in producing a good noodle.

20. What type of oriental noodle is fried?

SUGGESTED READING

BARONI, D. 1988. Manufacture of pasta products. Pages 191-216 in: Durum: Chemistry and Technology. G. Fabriani and C. Lintas, eds. Am. Assoc. Cereal Chem., St. Paul, MN.

DICK, J. W., and MATSUO, R. R. 1988. Durum wheat and pasta products. Pages 507-547 in: Wheat: Chemistry and Technology, 3rd ed. Vol. II. Y. Pomeranz, ed. Am. Assoc. Cereal Chem., St. Paul, MN.

HOSKINS, C. M. 1970. Macaroni products. Pages 246-299 in: Cereal Technology. S. A. Matz, ed. Avi Publishing Co., Westport, CT.

KONICK, C. M., MISKELLY, D. M., and GRAS, P. W. 1992. Contribution of starch and non-starch parameters to the eating quality of Japanese white salted noodles. J. Sci. Food Agric. 58:403-406.

MATSUO, R. R. 1975. Durum wheat—Production and processing. Pages 577-604 in: Grains and Oilseeds—Handling, Marketing, Processing, 2nd ed. V. Martens, ed. Can. Int. Grains Inst., Winnipeg.

MOSS, H. J., MISKELLY, D. M., and MOSS, R. 1986. The effect of alkaline conditions on the properties of wheat flour dough and Cantonese-style noodles. J. Cereal Sci. 4:261-268.

OH, N. H., SEIB, P. A., DEYOE, C. W., and WARD, A. B. 1983. Noodles. I. Measuring the textural characteristics of cooked noodles. Cereal Chem. 60:433-438.

Breakfast Cereals

CHAPTER 16

After fasting overnight, we must break the fast in the morning with—what else?—breakfast. Although certainly anything can be and is eaten at breakfast, morning is traditionally a time when many cereal products are consumed. Those products, in addition to sweet rolls and breads, can be divided into two types. The first is those that require cooking. They are as old as civilization. The second is the first of the convenience foods, the ready-to-eat cereals. They were developed only a little over 100 years ago by a group of vegetarians wanting to improve and add variety to their diets.

Cereals That Require Cooking

Cereals that are cooked before being served can be made from a number of grains. From wheat, for example, we obtain a farina. *Farina* is a fraction of middlings from a hard wheat. Soft wheat tends to give a product that becomes mushy after cooking. Farina can be obtained in about a 30% yield from hard wheat. Of course, the miller must find a use for the rest of the flour, which can be a problem after 30% of the prime middlings has been removed. The critical factor in consumer acceptance of farina is the particle size. U.S. standards specify that 100% of the product must pass a No. 20 sieve (833 μm) and not more than 10% must pass a No. 45 sieve (350 μm). Farina must be boiled in water for several minutes because, for the product to taste cooked, the particle must be wet throughout and the starch gelatinized.

An "instant" farina that is on the market cooks in about 1 min of boiling. This has been treated with proteolytic enzymes, which open avenues (capillaries) for water to penetrate the particles. Farina is usually enriched with vitamins and minerals, commonly added as a dry mix. The products are often flavored with malt or cocoa.

Most oats that are consumed directly as food are served as breakfast cereal. Of the type that must be cooked, the most popular by far is rolled oats. The process used to make rolled oats is given in Fig. 1. The cleaned oats are treated with dry steam at 100°C. This reduces the moisture content to about 6% and inactivates enzymes, particularly the lipase system. Oats are high in lipid and very susceptible to rancidity. Drying the hulls makes them more brittle and therefore easier to remove. The dehuller used to remove the oat hulls is similar to the rice huller discussed in Chapter 8.

The next step is to separate the hulls from the groats and the whole oats that were not dehulled in the process. In general, removing the hulls is not difficult; they are light and can be separated by aspiration. It is much more difficult to make a clean separation of groats from whole oats. Even a small percentage of whole oats remaining with the groats is unacceptable. Hulls in rolled oats are not palatable.

The groats are then rolled, or flattened, with large, very heavy rolls to give the rolled product. Flaked whole groats take 10–15 min to cook. The cooking time (i.e., the time required for the hot water to penetrate to the center of the flake) is determined by the thickness of the flake. Of course, the cooking time is controlled by the thickest point, and therefore a uniform piece is desirable. To obtain rolled oats that cook more quickly, one must produce a thinner flake. This is accomplished by cutting the groats into two or three pieces before flaking. The smaller pieces give a thinner flake that cooks faster (3–5 min), although the cooked cereal will not retain its quality if held under hot conditions such as on a steam table (the product becomes a gruel). Thus, we have the classic trade-off between cooking time and product stability. Newer products are available that cook just upon addition of boiling water. Examination shows that these are very thin, and such products have little or no stability to storage after cooking.

Corn is also used to produce a breakfast cereal that requires cooking. The product is called "grits" or occasionally "hominy grits" and is very popular in the southern part of the United States. It is made from white corn. The grits are produced by dry milling of corn and are essentially small pieces of endosperm. As with farina, the particle size is important. A similar breakfast cereal is also made from rice.

Ready-to-Eat Cereals

CORN FLAKES

For many years, the most popular ready-to-eat cereal has been corn flakes. Corn is dry milled to remove the germ and bran. In the process,

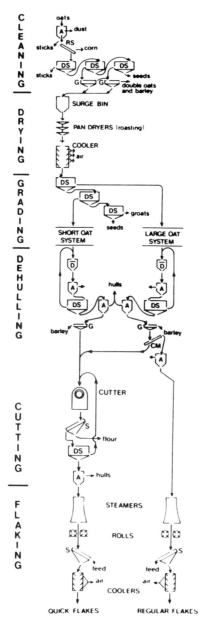

Fig. 1. Flow for producing rolled oats. A, aspirator; RS, receiving sieves; DS, disk separators; G, width graders; D, dehuller; CM, cell machine (gravity table); S, sieve.

the endosperm is usually split into two pieces. These large pieces (half of the kernel) are No. 4 or No. 5 grits, the starting material for corn flake manufacture. The large grits retain their identity throughout the process, each producing a single flake. Part of the corn endosperm is, of course, broken into smaller grits and therefore cannot be used in corn flake manufacture.

The corn grits are pressure cooked with a solution containing sugar, malt (nonenzymatic), and salt. Typical cooking conditions are 2 hr at 18 psi of steam pressure, but different lots of corn may vary considerably in cooking time. The end point of cooking can be determined by visual inspection. Uniform translucency is desirable because it indicates that the water has penetrated to the center of the piece.

After cooking, the lumps are broken up and the cooked grits partially dried. The drying is accomplished in a tower dryer in which the wet product falls countercurrent to a stream of hot air ($\sim$65°C). The height of the tower may be several floors. This process dries the outside of the particles so that they are no longer sticky. After cooking, the particles contain about 50% moisture, which is reduced in the dryer to about 20%. However, the moisture is not uniform. The piece is dry on the outside and moist in the interior. Therefore, it is given a temper time (about 24 hr) to allow the moisture to equilibrate

After being tempered, the grit is ready for flaking. The flaking rolls are large smooth rolls weighing as much as a ton each (Fig. 2). A pressure of about 40 tons is maintained at the point of contact. After coming from the flaking rolls, the flakes are toasted for about 50 sec at 300°C. The toasting not only dehydrates (<3% moisture) but also browns and blisters the product. After cooling, the flakes may be sprayed with a solution of vitamins and minerals.

WHEAT FLAKES

For wheat flakes, the whole kernel is used and each kernel makes one flake. The wheat is tempered, often with atmospheric pressure steam to raise the moisture content to about 21%. The steamed wheat is then bumped between smooth rolls. *Bumping* is passing the wheat between rolls set just slightly closer together than the thickness of the wheat. This causes the kernel to be distorted, at least temporarily. The distortion ruptures the bran, which is not as flexible as the other parts of the kernel. When intact, the bran acts as a container for the endosperm. As the endosperm takes up water, it swells. The bran limits the degree of swelling and thereby the amount of water taken up. Rupturing the bran allows the kernel to take up more moisture.

After being tempered, the kernels are cooked in a pressure cooker with sugar, salt, and malt flavor added. Normal cooking conditions are 90 min at about 20 psi. After cooking, the kernels are soft, translucent, and contain about 50% moisture. They are dried to about 21% moisture, tempered, and sent to the flaking rolls. The rolls are similar to those used for corn flakes. Just before the kernels go to the rolls, they are heated rapidly, usually by infrared lamps, to about 88°C; this *plasticizes* the kernel. If a wheat kernel is flattened with great force, it will tear and give an irregularly shaped piece (Fig. 3) that is fragile and will fall apart during shipping. Plasticizing the kernel allows it to flow under force and give a more uniform piece.

After leaving the rolls, the wheat flakes contain about 15% moisture. To obtain the desired crispness, they are oven toasted and dried to a moisture content of less than 3%.

Fig. 2. A set of flaking rolls.

SHREDDED BISCUITS

Shredded biscuits are one of the oldest of the ready-to-eat cereals. They are made from whole wheat with no addition of flavoring agents. The whole wheat is boiled at atmospheric pressure for 1 hr, which increases the moisture content to about 50%. The cooked kernels are tempered to equalize the moisture content and then sent to shredding rolls. The shredding rolls are about 8 in. in diameter and the length of a biscuit. They produce a set of wet strands of dough (ground whole-wheat paste). Eighteen to 20 pairs of rolls layer the dough strands lightly on top of each other, as illustrated in Fig. 4. The layered strands of dough are separated into biscuits by passing them below blunt knives.

The fragile biscuits are then baked at a relatively high temperature for 10–15 min. This toasts the outside but leaves the interior still wet. The temperature is then lowered to about 120°C for the remainder of the bake. The final product moisture is about 11%. Because the biscuits are made from whole wheat, rancid odors are developed during storage. Even small amounts of these odors will make the product unacceptable. Therefore, the product is packaged in breather-type boxes with no inner or outer gas barriers. This allows the rancid odors to diffuse from the boxes and gives the product a longer shelf life. A similar system is used with rolled oats but cannot be used for those cereals that are sensitive to moisture. Another system to increase the shelf life of cereals is to use package materials that contain antioxidants such as butylated hydroxyanisole (BHA) and butylated hydroxytoluene (BHT). This system is quite effective for increasing shelf life.

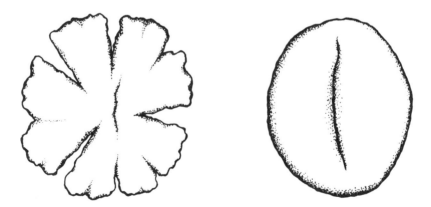

Fig. 3. Flattened kernel, without plasticizing (left) and with plasticizing (right).

CEREAL GRANULES

Cereal granules (as in Grape-Nuts) also are one of the oldest types of ready-to-eat cereals. They are made from a stiff (low moisture) dough prepared from wheat flour, barley flour, salt, yeast, and water. The dough is fermented at 27°C for about 5 hr and then made into loaves and baked without a proof period. The resulting loaves, which are quite dense, are fragmented by shredding knives. The fine particles are removed and reused in later doughs. The larger fragments are baked for an additional 2 hr or so at 120°C. The pieces are then reground and sized. Thus, Grape-Nuts are not nuts or grapes but very dense and hard pieces of toast in very small sizes.

PUFFED CEREALS

For the *puffing* of cereals, i.e., greatly decreasing their bulk density, two general methods are used. The first is the sudden application of heat at atmospheric pressure. In this technique, water is vaporized before it has time to diffuse to the surface of the piece. The internal vaporization then expands, or puffs, the product. The second type is the sudden transfer of a piece containing superheated water to a lower pressure, thus allowing the water to suddenly vaporize. Both types depend upon water going to a vapor as the driving force. The key

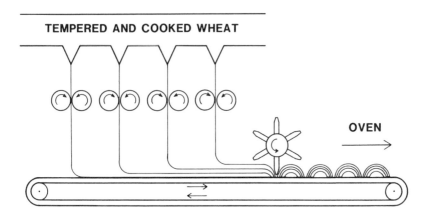

Fig. 4. Scheme for making shredded biscuits.

to the degree of puffing is the sudden change in temperature or pressure. Examples of processes using these techniques are given below.

Oven-puffed rice is an example of the sudden application of heat. Milled rice is cooked at 15 psi until it is uniformly translucent, dried to 30% moisture, tempered for 24 hr, and dried to 20% moisture. The intact kernels are subjected to radiant heat to plasticize the outside of the kernel. The kernels are bumped (to destroy their internal structure) and again tempered for 24 hr. Then they are passed through the oven at 300°C for about 30 sec. In this technique, the expansion is not great, just two to five times.

Pressure-puffed (often referred to as gun-puffed) products can be made from a dough mixed and steam-cooked to about 40% moisture. The dough is forced through an extruder to give it a specific shape and is dried to 12–15% moisture. The pellets are then loaded into guns or popping vessels. These guns are about 30 in. long and 6 in. in diameter. They are sealed, rotated, and heated to 425°C. Pressure may build to 200 psi. The trip valve is opened and the material explodes out. Many variations are possible, using different flavors, shapes, types of doughs, etc. Wheat and rice kernels can also be gun-puffed. Milled rice and pearled wheat are generally used, as the bran is blown off during the process and would be unsightly in the product. The degree of expansion obtained with gun puffing (15–20 times) is much greater than with oven puffing.

Puffed products must be kept at less than 3% moisture to maintain crispness. The more the product is expanded, the more critical and harder the levels are to maintain. Thus, gun-puffed products require special packaging.

Extrusion cooking is another method of expanding cereal products. It is a continuous process and uses both temperature and pressure to expand the kernels. In general, a cereal flour or grit that has been moistened with steam or water is fed into the extruder. An outline of the barrel of a double-screw cooker-extruder is given in Fig. 5. The screw's channel depth decreases at the end close to the die. The cooker-extruder is essentially a nonefficient screw pump. Heat is generated by friction, and the barrel may be heated by steam. Under the temperature and pressure, the cereal product melts to form a plastic mass in the extruder barrel. The temperature may approach 200°C and the pressure 500 psi at the die head. Under these conditions, the dough is quite flexible and will adapt to any die configuration. Upon exiting the die, the dough expands as the pressure is released. Moisture is flashed off and, of course, this cools the product. Feed material generally contains 12–20% moisture and the final product 8–15% moisture.

Therefore, most of the products must be dried after extrusion.

The type of cereal flour, its composition (particularly the level of fat), and many other things affect the extrudate and its degree of expansion. Monoglycerides have been used to produce a smoother product that is more uniform. However, they also reduce the product's expansion. The major advantage of the cooker-extruder is that it can handle cereal flours at a relatively low moisture content. In most systems, more water must be added, and later in the system that water must be removed, which is very expensive. The cooker-extruder has a great future because of these economic advantages.

COATINGS

Many cereal products are sugarcoated before they are sold. Many pseudoexperts in nutrition are convinced that we are forcing sugar upon an uninformed population. Actually, the sugarcoating is for quite a different reason. It protects the product from moisture and thus gives longer shelf life for certain products. There appears to be little difference between adding sugar at the point of manufacture or at the point of consumption.

The coating process is quite simple. A cement mixer type of apparatus is used to keep the cereal agitated while the molten sugar syrup is slowly dripped onto the mass. Coconut oil is often added to decrease foaming and to keep the particles separate. The process is like making popcorn balls at home.

The syrup hardens quite rapidly upon cooling. The glaze accounts for 25-50% of the product weight, mainly because of its very high density compared to that of the cereal product.

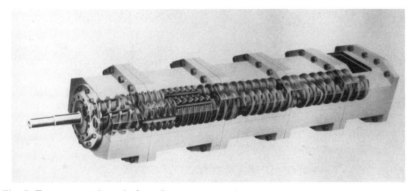

Fig. 5. Transparent barrel of a twin-screw extruder.

REVIEW QUESTIONS

1. What is farina?
2. What are the U.S. standards for particle size of farina?
3. How are instant-cooking cereals prepared?
4. Why are oats heated with steam before processing?
5. How are quick-cooking rolled oats produced?
6. What is the major disadvantage of quick-cooking rolled oats?
7. What are grits?
8. Outline the process of making corn flakes.
9. Why are wheat kernels plasticized during the making of wheat flakes?
10. How are shredded biscuits made?
11. How are shredded biscuits packaged and why?
12. Explain how cereals are puffed.
13. Explain how the cooker-extruder works.
14. Why are some ready-to-eat cereals coated with a sugar coating?

SUGGESTED READING

BROCKINGTON, S. F., and KELLY, V. J. 1972. Rice breakfast cereals and infant foods. Pages 400-418 in: Rice: Chemistry and Technology, 1st ed. D. F. Houston, ed. Am. Assoc. Cereal Chem., St. Paul, MN.

FAST, R. B., and CALDWELL, E. F., eds. 1990. Breakfast Cereals and How They Are Made. Am. Assoc. Cereal Chem., St. Paul, MN.

LINKO, P., COLONNA, P., and MERCIER, C. 1981. High-temperature, short-time extrusion cooking. Pages 145-235 in: Advances in Cereal Science and Technology, Vol. 4. Y. Pomeranz, ed. Am. Assoc. Cereal Chem., St. Paul, MN.

LOVING, H. J., and BRENNEIS, L. J. 1981. Soft wheat uses in the United States. Pages 169-207 in: Soft Wheat: Production, Breeding, Milling and Uses. W. T. Yamazaki and C. T. Greenwood, eds. Am. Assoc. Cereal Chem., St. Paul, MN.

MATZ, S. A. 1970. Manufacture of breakfast cereals. Pages 221-245 in: Cereal Technology. S. A. Matz, ed. Avi Publishing Co., Westport, CT.

YOUNGS, V. L., PETERSON, D. M., and BROWN, C. M. 1982. Pages 49-105 in: Advances in Cereal Science and Technology, Vol. 5. Y. Pomeranz, ed. Am. Assoc. Cereal Chem., St. Paul, MN.

Snack Foods

CHAPTER 17

One way to define snack foods is to say that they are generally eaten just as they are removed from the package. Such a definition would include cookies and crackers, and even bread and breakfast cereals. Perhaps we can use this definition with the understanding that there are some notable exceptions. The fact that one has a problem finding a neat definition may also explain why the line between snack food manufacturers and other food manufacturers has become fuzzy in recent years.

Corn Products

Popcorn is, of course, the original snack food, one that has been in use for many centuries. Not only is popcorn our oldest cereal snack, but it is, in many ways, unrivaled by many later snacks. It has a delicate corn flavor and a delightful texture. It can easily be flavored with fat and salt or caramel, cheese, and many other flavors. Of course, the purist would want only fat and salt.

The quality of popped corn is determined by such factors as popped volume, shape of the popped kernels, tenderness, and, of course, flavor. During popping, the volume of the corn increases up to about 30 times. The popped volume is of particular importance because popped corn is bought by weight but generally sold by volume. The tenderness of the popped corn has also been positively correlated with popping volume.

Among the cereal grains, only popcorn, and certain lines of sorghum and pearl millet, pop. All grains expand in volume to lesser extents, but only these three give the explosive pop with the accompanying large increase in volume. This raises an interesting question: Why does popcorn pop? The obvious answer is that the moisture in the kernel vaporizes and expands the kernel. Although this is undoubtedly true, it is only part of the answer, because the other cereals also contain moisture but do not pop. The pericarp of popcorn acts as a pressure vessel, allowing the water in the kernel to be superheated. At a certain temperature, the pressure becomes so large that the pericarp ruptures, allowing the endosperm to expand. With popcorn, the failure of the pericarp occurs at about 177°C.

In addition to serving as a pressure vessel, the pericarp is also an important quality factor. It is the "hull" that all popcorn consumers know but few enjoy between their teeth. The effect of the intactness of the pericarp on popped volume is shown in Table I.

A longitudinal cross section of an unpopped popcorn kernel is shown in Fig. 1. As with all grains, the kernel is made up of pericarp, endosperm, and germ. In corn, the endosperm is composed of a translucent portion and an opaque portion. The degree of expansion in popcorn has been shown to be related to the amount of translucent endosperm. This endosperm is tightly packed and contains no air spaces, whereas the opaque endosperm has many air spaces (see Chapter 1).

During popping, the pericarp (Fig. 2) appears to undergo no change other than the obvious fracture. It is often blown free or partially free of the endosperm. The aleurone and subaleurone cells also appear to undergo minimal change during popping. At the outer edge of the kernel, moisture loss is excessive. Deeper in the endosperm, the expanded structure becomes more evident. The starch is greatly expanded, but the protein bodies are still visible and apparently have not been changed by the popping process. As shown in Fig. 3, the opaque endosperm acts much differently than does the translucent endosperm during popping. The starch granules in the opaque endosperm appear to stay intact and do not expand during popping. The moisture vaporizes into the void surrounding the granules, creating large voids. In the trans-

TABLE I
Effect of Pericarp Treatments on Popping Volume[a]

Treatment	Popped Volume ($cm^3/10$ g of grain)
Control, intact kernels	270
Pericarp cut just through	120
Pericarp cut just through in two places	90
Opaque endosperm cut into	70
Kernel cut in half	40
Pericarp removed by hand	20
Unpopped corn	11

[a] From R. C. Hoseney, K. Zeleznak, and A. Abdelrahman. 1983. J. Cereal Sci. 1:43-52. Used by permission.

lucent endosperm, the moisture vaporizes into the hila of the starch granules (Fig. 3) and expands the granules. The germ does not appear to be affected by popping.

Corn-nuts, another corn snack, are made from a specific corn type that has very large, opaque, white kernels. These are tempered and then fried in fat and salted. They have a very nice taste but are rather hard. Care must be taken or teeth can be broken with these snacks. A similar product from wheat has not gained equivalent popularity.

In India, certain varieties of sorghum and pearl millet are popped and eaten much as popcorn is eaten in North America. The major difference, in addition to the flavor and size, is that, in India, the grains

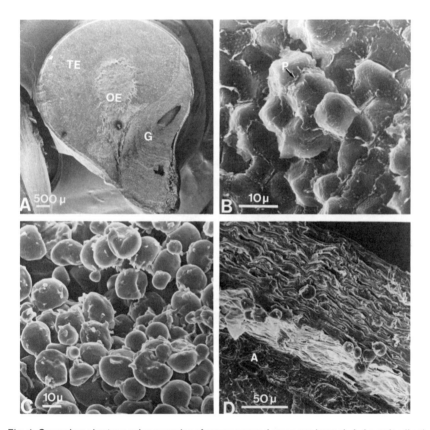

Fig. 1. Scanning electron micrographs of an unpopped popcorn kernel. A, Longitudinal section, showing translucent endosperm (TE), opaque endosperm (OE), and the germ (G); B, translucent endosperm, showing starch and protein bodies (P); C, opaque endosperm; D, outer edge of the kernel, showing the aleurone (A) and the pericarp.

are popped in hot sand rather than in oil. The fine sand can then be removed (at least partially) from the popped kernel.

Corn curls are another corn-based snack. They are made from corn flour with a cooker-extruder. (The instrument is discussed in Chapter 16.) The expansion obtained in the extruder is somewhat less than that obtained with popcorn (12 times as opposed to 30 times). Other than that, the procedures are similar; the extruder, rather than the pericarp, holds the pressure. Corn curls are often flavored by being sprayed with oil and then coated with cheese or other flavoring agents. In newer types of snacks, also produced with the cooker-extruder, the product is not expanded after cooking but is cooled and cut into small

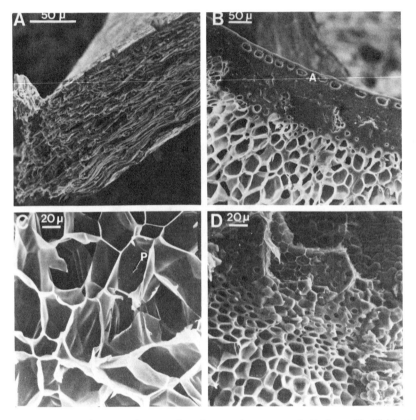

Fig. 2. Scanning electron micrographs of popped popcorn. A, Pericarp "hull" blown free as a result of popping; B, outer edge of the endosperm, showing the aleurone (A); C, inner portion of the endosperm, showing the protein bodies (P) still intact; D, the endosperm in a partly popped kernel, showing starch granules ranging from intact to expanded forms.

shapes. These products are then expanded by frying or hot-air popping. They have the advantage of having a fresh taste.

Masa and Its Products

Corn, or maize, has been used directly as a human food for centuries in South and Central America. In practically all of its uses, the corn is first made into *masa*. The traditional production method that is still practiced in many areas of Latin America and the industrial method of making masa in the United States are not very different. Corn is heated with water containing calcium hydroxide (about 0.5–1.0%) or

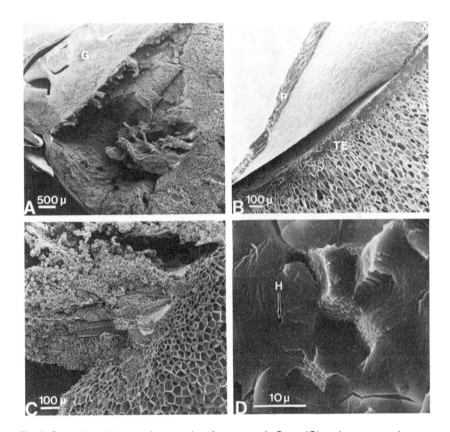

Fig. 3. Scanning electron micrographs of popcorn. A, Germ (G) and opaque endosperm of a popped kernel; B, pericarp (P) separating from the translucent endosperm (TE) of a popped kernel; C, interface between the opaque and translucent endosperm in a popped kernel; D, hilum (H) at the center of a broken starch granule from an unpopped kernel.

with wood ashes if calcium hydroxide is not available (Fig. 4). The temperature is brought to at least 82°C but is generally held below boiling for a relatively short time (<1 hr). The corn is then placed in stoneware jars or in bulk to cool slowly to room temperature (generally overnight). This produces *nixtamal* (heat-treated, alkaline water-steeped corn), which is then washed to remove "hulls" and excess calcium hydroxide and then ground either by hand or machine with stone burrs. The resulting wet-ground product is masa. It contains about 55% moisture and forms a slightly cohesive dough, presumably because of its particle size distribution and moisture.

During the cooking of the corn, relatively little of the starch is gelatinized. The temperature is sufficiently high to gelatinize the starch,

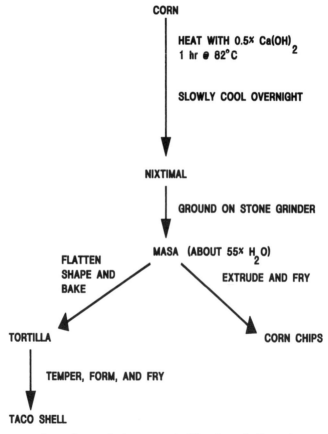

Fig. 4. Process scheme for producing masa, tortillas, taco shells, and corn chips.

but the amount of water is limited. In addition, the soluble salts and sugar in the kernel increase the starch gelatinization temperature so that most of the starch is not gelatinized. The corn kernel is quite hard and dense; thus, a long period of time is needed for the water to diffuse to the center of the kernel. This is apparently the reason for the long soak (steep) time at elevated temperatures. The calcium hydroxide is added for two reasons. First, it gives a flavor that is quite compatible with corn and one that we identify with corn products. Second, the alkaline character of the calcium hydroxide weakens the outer layer of the corn pericarp. After the cooking, this can be removed by rubbing the corn or even by washing it thoroughly. In some products, the pericarp is retained (after all, it increases the yield, even though it gives small dark specks); in others it is washed away before grinding. The cooking water is removed before the product is ground. Small amounts of water may be added during grinding to obtain the desired moisture content. The amount of cooking required varies from corn to corn and is apparently related to the rate of water penetration. It is important that all of the kernel be hydrated to the same extent. Therefore, determination of the end of cooking is an important aspect of the process. The grinding also appears to be an important step. Stone mills are generally conceded to be necessary to produce good masa. This is presumably related to the particle size distribution that is produced.

Masa is the basic material for producing a number of products. It is pressed, either by hand or mechanically, into a round, flat pancake shape and baked to form a tortilla. This is the basic flat bread that is made from either wheat flour or corn masa flour. The baking is generally at a relatively high temperature, but not high enough to permanently puff the product (they often puff during baking and then collapse), and generally the tortilla is turned two or three times during baking. Masa can also be extruded from a forming extruder into hot fat to produce a corn chip (Fig. 4). Other types of corn chips as well as taco shells are made by frying a tempered tortilla in deep fat. This gives quite a different texture. The tempering, letting the tortilla set under conditions that do not allow it to dry, is to allow the moisture to equilibrate in its interior. The amount of moisture in the product when it is fried is an important determinant of the texture of the product.

Cohesive Properties of Nonwheat (Nongluten) Doughs

Masa, as well as the *chapatti* or *roti* doughs made from sorghum and pearl millet flours, is an example of a dough having cohesive properties but not containing gluten. Scattered references in the technical and patent literature refer to the development of corn or sorghum proteins into cohesive doughs. However, none of the reports is very convincing.

These doughs are unquestionably cohesive. However, they do not have the elastic properties of a wheat flour dough. Thus, we are correct in calling the doughs cohesive rather than elastic. The cohesive property can probably be explained in the same way as we would explain the cohesive properties of "mud pies"; inert particles are held together by the surface tension properties of water. When the voids between the particles are just filled with water, the system is most cohesive. Such an explanation demands that the systems (doughs) be sensitive to the amount of water in the system and to the particle size of the flours or masa used. This has been demonstrated. The secret to producing a cohesive nonwheat dough is to produce a fine particle size and add the correct (optimum) amount of water.

Synthetic Nuts

An interesting by-product of designing foods for astronauts is the synthetic nuts made with wheat germ as a major ingredient. For such products, an oil-in-water emulsion is produced, with a film-forming protein (from wheat germ) in the aqueous phase. The emulsion is then carefully dried to set the protein into a solid. The products can be produced with the texture of nuts. Because the oil is protected from the outside of the pieces, and thereby from the air, the products have a long shelf life. This is a classic example of using Mother Nature's trick of compartmentalization to obtain good shelf life.

Wheat germ is high in both fat and enzymes. As long as it is intact, it is relatively stable; however, during milling, the germ is flattened and then the fat and enzyme are mixed. As a result of that mixing,

TABLE II
Typical Formula for Pretzels

Ingredient	Parts[a]
Flour	100
Shortening	1.25
Malt syrup (nondiastatic)	1.25
Yeast	0.25
NH_4HCO_3	0.04
Water (variable)	about 42

[a] Based on flour.

the germ rapidly becomes rancid. To stabilize the germ, it is roasted or dry heated to denature the enzyme.

Pretzels

Pretzels are a baked food that is unique both in shape and in having a hard outer surface. A typical formula for pretzels is given in Table II.

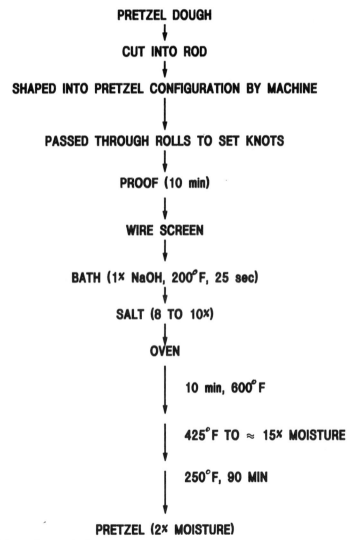

Fig. 5. Scheme for producing pretzels.

The formula is somewhat unusual in that it contains both yeast and a chemical leavening agent. Occasionally, one finds a formula that contains only chemical leavening. The water level is about two thirds of the amount that would be used with a bread dough. Thus, the dough is dry and tough.

The pretzel-making process is outlined in Figs. 5 and 6. A dough is mixed and then rolled into a rope. It goes through a twisting machine that forms it into the traditional shape. The dough then passes through rolls to set the knots. The fermentation given the dough after mixing and before processing is generally quite short (30 min or so). This is very little fermentation considering the low level of both yeast and water. After passing through the twister, the dough is generally given a relaxation period or proof of 10 min or so. The proofed pieces are then transferred to a wire mesh band and conveyed through a bath of hot 1% sodium hydroxide. The total time in the lye is about 25 sec. The bath temperature is about 93°C. The bath gelatinizes the starch on the outer surface of the piece. Immediately after the bath, the pieces are "salted" with a coarsely flaked salt (about 2% of the final product). After salting, the pieces are baked, starting with a high heat to caramelize the gelatinized starch on the surface and produce a dark brown color on the product. The temperature later in the oven is lower than at the front of the oven to allow moisture to come out of the dense knot areas. If heated too fast, the product will check and fall apart during shipping.

A recent trend in the industry is to use sticks or other shapes in place of the traditional shape (Fig. 7). This eliminates the knots, and the product can be baked at a higher temperature and for a much

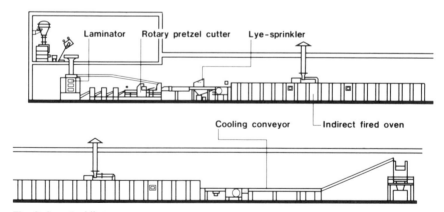

Fig. 6. A pretzel line.

shorter time. Machines have also been developed to extrude dough in the traditional shape so that the pieces can be wire-cut. Because all the dough is the same thickness, it also can be baked faster.

An important question about the pretzel-making process is: What happens to the sodium hydroxide? Clearly, we do not want to eat lye. If you remember your freshman chemistry, sodium hydroxide reacts with the carbon dioxide in the air to form sodium bicarbonate.

In the larger U.S. cities and throughout Europe, one finds large, soft pretzels that are quite different from those described above. These are similar in formulation but are not taken through the lye bath. Some are bathed with sodium carbonate and others are just baked in a hot oven.

The flour used for pretzels appears to vary widely, depending upon where the plant is located and also probably upon the processing equipment. Generally it is a soft wheat flour of 9.0–9.5% protein.

Bagels

Bagels originated in Vienna. After horsemen were responsible for driving an attacking force of Ottoman Turks from the city, the bakers wanted to produce a new bread in the riders' honor. They attempted to form the bread in the shape of a stirrup. To have it retain that

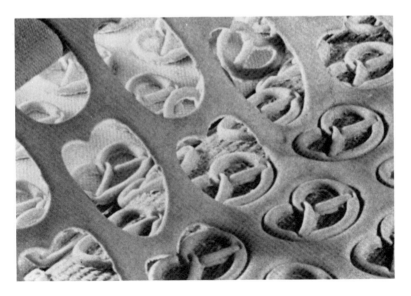

Fig. 7. Dough web with pretzels cut out and the scrap web being lifted away.

Fig. 8. Effect of retardation time on the external appearance of bagels. Retardation times: 10 hr (left), 18 hr (center), 30 hr (right).

shape, they made a strong dough by using high-protein flour with relatively low water absorption. There was still a problem, as the stirrup shape was lost during baking. The problem was solved by boiling the dough before it was baked, which gave it the necessary strength to retain its shape. It also made a very unique baked product.

Bagels in the United States were an ethnic food of the Northeast until fairly recently. Now bagles can be found in essentially all parts of the country.

The traditional process for bagel production is to mix a stiff dough with a small amount of yeast, sugar, and salt. The dough is formed into the desired shape and retarded (refrigerated) at 1°C for 18 hr. It is removed from the retarder, allowed to warm at room temperature for 15 min, and then placed in boiling water for 2 min. After being turned, it is boiled for an additional 2 min. The dough is allowed to drain for 30 sec and is then baked.

The retardation step appears to be necessary to obtain the hard, shiny surface characteristic of traditional bagels (Fig. 8). Many of today's bagels are made by modified systems, with short or no retardation times and often steaming rather than boiling in water. The newer systems are much more economical but do not give a product of the same quality.

The flour for bagels is usually a high-protein flour. If not, the formula usually contains added wheat gluten.

REVIEW QUESTIONS

1. Why are snack foods difficult to define?

2. What determines the quality of popped corn?
3. Explain why popcorn pops.
4. Why is the ratio of translucent to opaque endosperm important in popcorn?
5. What happens to the germ as a result of popping corn?
6. What is masa and how is it made?
7. What is nixtamal?
8. How is masa used to make tortillas, corn chips, and taco shells?
9. What gives nonwheat doughs their cohesive properties?
10. How are synthetic nuts made?
11. How is wheat germ stabilized?
12. What is a typical formula for pretzels?
13. What is responsible for the hard dark glaze on the outside of pretzels?
14. How is the sodium hydroxide removed from the products?
15. What is the difference between hard and soft pretzels?
16. What is responsible for the hard, shiny surface on a bagel?
17. What type of flour is used for bagels?

SUGGESTED READING

ALLEN, J. C., and HAMILTON, R. J. 1984. Rancidity in Foods. Elsevier, New York.

LINKO, P., COLONNA, P., and MERCIER, C. 1981. High-temperature, short-time extrusion cooking. Pages 145-235 in: Advances in Cereal Science and Technology, Vol. 4. Y. Pomeranz, ed. Am. Assoc. Cereal Chem., St. Paul, MN.

MATZ, S. A. 1984. Snack Food Technology, 2nd ed. Avi Publishing Co., Westport, CT.

MERCIER, C., LINKO, P., and HARPER, J. M., eds. 1989. Extrusion Cooking. Am. Assoc. Cereal Chem., St. Paul, MN.

SERNA-SALDIVAR, S. O., GOMEZ, M. H., and ROONEY, L. W. 1990. Technology, chemistry, and nutritional value of alkaline-cooked corn products. Pages 243-307 in: Advances in Cereal Science and Technology, Vol. 10. Y. Pomeranz, ed. Am. Assoc. Cereal Chem., St. Paul, MN.

VICKERS, Z. M. 1988. Crispness of cereals. Pages 1-19 in: Advances in Cereal Science and Technology, Vol. 9. Y. Pomeranz, ed. Am. Assoc. Cereal Chem., St. Paul, MN.

Feeds

CHAPTER 18

In many ways, the distinction between food and feed is somewhat arbitrary. Many of the materials that we use as food also end up being used as feed. It is also true that we can view feed as food that requires just one more processing step—a step performed by the animal to which we feed the material. You might ask if this applies to pet feed as well as to feed for the domestic animals that we use for food. The answer is: of course, but we must consider it as food for the spirit and not for the body.

Of course, it is possible to feed animals the cereal grains with no processing. In fact, this used to be common practice in years gone by. Cattle were fed whole corn or corn that had been cracked into a few pieces, and sorghum was fed whole. A stroll through a barnyard where this was the practice, however, would convince one that there was a problem with digestibility. Whole kernels of corn or even cracked kernels would pass through the cattle's digestive system essentially untouched. To make a more efficient conversion of the feed, the farmer would run hogs in the same barn lot. Behind the hogs, chickens were allowed to roam. This was not a bad system, but it did require that the correct number of each animal be available. Also, hogs had a nasty habit of occasionally eating one of the chickens and thus throwing the whole system out of balance. On today's farms, the farmer does not keep all those animals but instead concentrates on one species. Farmers also tend to keep a large number of whatever animal is being raised. Clearly, our "mother nature" system needs a hand to be efficient.

Grinding

COARSE GRINDING

Anything used to break or remove the hull or outer layer of cereal grains generally makes an improvement in *feed efficiency* (the amount of gain divided by the feed consumed). The outer layers of the grain are there to give protection, including protection against digestion by animals that eat them. The hulls that are removed may be very useful as fiber in the diets of *monogastric* animals. However, even with *ruminant* animals, the hull, while digestible, slows the rate at which the remainder of the grain is digested.

A number of processes can be used to increase the digestibility of the grain. Passing the grain through a roller mill with rolls set close enough to break the grain is effective in breaking the outer layers and exposing more of the endosperm to digestion. The passage thorough the roller can be done dry (*dry rolling*), with the grain at its native moisture content, or after the grain has been treated with moisture. Dry processing tends to create many fine particles and therefore a dusty product.

Addition of moisture before the grinding reduces the dust. The moisture addition can be as water or as steam (*steam rolling* or *crimping*). The steam treatment is usually short, 1-8 min, and appears to have little, if any, effect on the digestibility of the grain over that achieved by dry rolling. However, it does improve palatability.

FINE GRINDING

Processing the feed material by grinding it fine is a big step in the direction of improved feed efficiency. Fine grinding not only breaks the pericarp, mother nature's protective device, but also greatly increases the surface area of the material. If we increase the surface area, then more of the feed material is available for the digestive enzymes to attack.

The grinding of feed material should be as efficient as possible, that is, grinding the most tons with the least horsepower. In feed grinding, this usually means using a *hammer mill* (Fig. 1). In a hammer mill, the hammers are affixed to a rotating shaft, and the grinding action is essentially by impact. The material to be ground is retained in the grinding area by a screen. The screen has holes in it to allow material of sufficiently small size to pass through. Larger material is retained on the screen to be ground again. By varying the size of the screen, we can roughly control the particle size of the ground material. Of course, the load or rate of feed also affects the grinding efficiency. The particle size distribution resulting from hammer milling is usually quite wide. The hammer mill is a low-cost, efficient grinder.

Although fine grinding has many advantages, it also has some disadvantages. The major ones are that fine grinding is relatively expensive and that it produces a very dusty product. Another disadvantage is that the fine material can easily be blown out of feed bunks and is also subject to other forms of waste. The fine, dusty material also makes the feed less palatable to certain animals.

Pelleting

The problems with fine mash were solved by pelleting the feed into more dense material. This is conveniently done with a pellet mill such as that shown in Fig. 2. The feed material, along with water (usually in the form of steam), is fed into the pellet mill. The material is forced into the pellet die with great force. After coming through the die, the pellets are broken off. They are hard, but much less hard than the original corn or sorghum. They can be easily handled with relatively little loss or waste and can be made in various sizes and shapes, from large range cubes designed to be fed on the ground to range cattle to small pellets for songbirds.

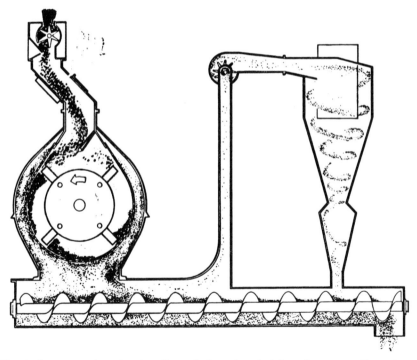

Fig. 1. Cut-away view showing flow of air and material as they move through an air-assisted hammer milling system.

It is not totally clear what forces hold the pellets together. However, pure starch granules form a pellet when subjected to sufficient pressure. Apparently, when the starch granules are forced close enough together, short-range forces are sufficient to aggregate them. Presumably, similar forces are at work with the feed material. Studies have also shown that different materials in the feed make the pellets stronger or weaker. Examination of the pellets shows them to be hard and glassy on the surface but mealy in the center. If the pellet were glassy throughout, we would have defeated our purpose for grinding the feed material in the first place. We need the pellet to be hard enough to be stable but weak enough to break apart when the animal eats it. Scanning electron microscopic studies of the pellets show the inside to be porous

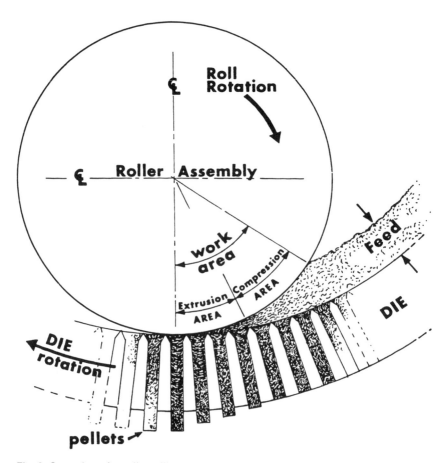

Fig. 2. Operation of a pellet mill.

and the outer layer to be dense (Fig. 3). Differential scanning calorimetric studies have shown that most of the starch on the outside of the pellet has been gelatinized (lost its crystallinity). This might be expected, considering the shear and heat it has undergone in the pellet die. These studies also showed that the starch in the center of the pellet was not gelatinized.

Feed Formulation

With the advent of fine grinding and then of pelleting the feed material came the possibility of mixing various grains and other feed materials to produce what was needed rather than using what mother nature put into the corn or sorghum. We could now use various mixtures and formulate the feed to meet the specific needs of a type of animal.

Coupled with the above was the need to accurately know the requirements of the various animals. As this information was developed, feeds were formulated to improve the efficiency of the feed and to lower the cost of each pound of gain.

Alternatives to Grinding

DRY HEAT PROCESSING

Micronizing, popping, and *roasting* are three processes used to dry-heat grain for animal feed. Micronizing is the heating of grain with infrared heaters. The moisture content is reduced to about 7%, and the grain is slightly expanded. Micronized corn generally has a bulk density of about 25 lb/ft^3 (0.40 g/cm^3) compared to the 45 lb/ft^3 (0.72 g/cm^3) for untreated corn. Micronizing improves both the rate of gain and the grain's feed efficiency.

Popping is achieved by rapid, intense heating of the grain. Slow heating allows the water to escape, but rapid heating makes the water expand all at once and thereby expand the product. The degree of expansion depends on the heating rate and the type of grain being heated. Most feed grain does not pop in the sense that popcorn pops. However, most cereal grains do expand under similar treatment. As a result of the rapid intense heating, not only does the grain expand, but also the starch is gelatinized. This results in the grain being much more available to digestive enzymes and/or organisms. Popped grain is a good material with which to start cattle on feed. However, its greatly reduced bulk density can be a problem, which manifests itself in two major ways. First, the feed is often blown from the feed bunks, causing waste. Second, the animal cannot consume sufficient feed, which results in reduced gains. Most users of popped cereals roll the popped grain in an effort to reduce its bulk density. Moisture is also often added, often in the form of molasses.

Roasted grain is similar to popped grain except that the grain is heated at a much slower rate. This allows water to be lost and therefore produces much less expansion. Depending upon the rate of heating, the starch may be gelatinized and a small amount of puffing may occur. This also leads to improved rates of gain and improved feed efficiencies.

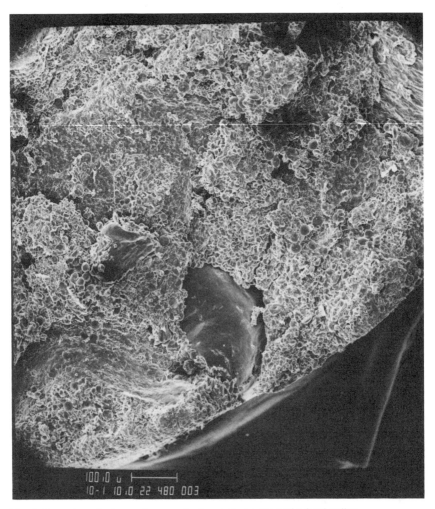

Fig. 3. Scanning electron micrograph of a cross-section of a feed pellet.

WET HEAT PROCESSING

The most popular wet processing of feed is *steam flaking*. Steam flaking differs from steam rolling in the amount of time used to steam the grain. In flaking, the grain is steamed for longer times, sufficient to gelatinize part of the starch. Gelatinized starch is much more available to enzymes than is raw starch. Thus, steam flaking has been shown to improve rates of gain and feed efficiency.

Extrusion cooking is another type of wet processing. In a cooker extruder, the grain is heated with water to a temperature and pressure that cause the grain to become a plastic material. This material is extruded to the atmosphere, where it expands. The expansion is caused by the super-heated water, which is flashed off as the material expands. The resulting material is glassy in texture but honeycombed with voids. The voids reduce the bulk density, and the glassy nature of the exudate makes it adsorb water slowly. The combination of low bulk density and slow adsorption make this feed useful for fish and other aquatic dwellers.

Specialty Feeds

The major specialty feeds encountered are various pet feeds, including those for animals in a zoo and the various fish and crustacean feeds. As discussed above, we often want fish feeds to float on water. The advantage here is that the farmer can see whether the fish are feeding and also determine whether they are being overfed. Too much unconsumed feed quickly leads to water quality problems. Crustaceans tend to be slow eaters, and the time the piece stays together is of great importance. Fish feeds need to be high in protein and relatively low in energy compared to feeds for other animals. Fish do not need energy to maintain their body temperature.

The object in pet and zoo feeding is different than that for most farm animals. Farm animals are raised for the milk or meat that they produce. How fast they put on weight or the amount of eggs or milk produced per unit of feed can be readily calculated. On the other hand, with pets the idea is for them to live a long and healthy life. This is somewhat harder to measure. In addition, there is the pet owner, who is, after all, the person making the buying decision. Most pet owners want their pet to eat the feed, even though the animal expended no energy lying in front of a warm fire all day. Also of importance is how the animal deals with the feed. It should not make a mess, and if it is an indoor animal, it should not give off a bad-smelling gas.

REVIEW QUESTIONS

1. What is the difference between food and feed?
2. How is feed efficiency calculated?
3. What is the difference between dry rolling and crimping?
4. What is a major problem with dry-rolled feed?
5. How would you describe the grinding action of a hammer mill?
6. What are some of the advantages of fine grinding?
7. What are some of the disadvantages of fine grinding?
8. What are range cubes and how are they used?
9. Explain how a pellet mill works.
10. What are three dry-heat processes used for feeds?
11. Why do fish need a lower-energy ration?

SUGGESTED READING

Feed Manufacturing Handbook, 3rd ed. 1985. Robert McEllhiney, ed. American Feed Manufacturers Association, Arlington, VA.

Figure Credits

CHAPTER 1

Figs. 1 and 4. Reprinted, with permission, from: M. M. MacMasters, J. J. C. Hinton, and D. Bradbury. 1971. Pages 51-113 in: Wheat Chemistry and Technology, 2nd ed. Y. Pomeranz, ed. Am. Assoc. Cereal Chem., St. Paul, MN.

Fig. 2. Courtesy of Gary Fulcher.

Fig. 3. Courtesy of the Wheat Flour Institute, Washington, DC.

Figs. 5, 6, 9-11. Reprinted, with permission, from: R. C. Hoseney and P. A. Seib. 1973. Bakers Dig. 47(6):26-28, 56.

Fig. 12. Courtesy of Corn Refiners Association, Inc., Washington, DC.

Figs. 13-16. Reprinted, with permission, from: J. L. Robutti, R. C. Hoseney, and C. E. Wassom. 1974. Cereal Chem. 51:173-180.

Fig. 17. Reprinted, with permission, from: D. B. Bechtel and Y. Pomeranz. 1980. Pages 73-113 in: Advances in Cereal Science and Technology, Vol. 3. Y. Pomeranz, ed. Am. Assoc. Cereal Chem., St. Paul, MN.

Fig. 21. Reprinted, with permission, from G. H. Palmer and G. N. Bathgate. 1976. Pages 237-324 in: Advances in Cereal Science and Technology, Vol. 1. Y. Pomeranz, ed. Am. Assoc. Cereal Chem., St. Paul, MN.

Fig. 29. Reprinted, with permission, from: L. Rooney. 1973. Pages 316-342 in: Industrial Uses of Cereals. Y. Pomeranz, ed. Am. Assoc. Cereal Chem., St. Paul, MN.

Figs. 30, 32, and 33. Reprinted, with permission, from: R. C. Hoseney, A. B. Davis, and L. H. Harbers. 1974. Cereal Chem. 51:552-558.

Figs. 34 and 35. Reprinted, with permission, from: R. C. Hoseney, E. Varriano-Marston, and D. A. V. Dendy. 1981. Pages 71-144 in: Advances in Cereal Science and Technology, Vol. 4. Y. Pomeranz, ed. Am. Assoc. Cereal Chem., St. Paul, MN.

Figs. 36 and 37. Reprinted, with permission, from: S. M. Badi, R. C. Hoseney, and A. J. Casady. 1976. Cereal Chem. 53:478-487.

CHAPTER 2

Fig. 2. Courtesy of Nancy Martin.

Fig. 3. Courtesy of Kathy Zeleznak and J.-L. Jane.

Fig. 4. Courtesy of Nancy Martin.

Fig. 5. Reprinted, with permission, from: K. Ghiasi and R. C. Hoseney. 1981. Starch/Staerke 33:428-430.

Fig. 6. Reprinted, with permission, from: T. L.-G. Carlson, K. Larsson, N. Dinh-Nguyen, and N. Krog. 1979. Starch/Staerke 31:222-224.

Fig. 7. Courtesy of D. Bath.

Fig. 8. Reprinted, with permission, from: D. French. 1984. Pages 184-247 in: Starch Chemistry and Technology, 2nd ed. R. L. Whistler, J. N. BeMiller, and E. F. Paschall, eds. Academic Press, Orlando.

Fig. 12. Reprinted, with permission, from: J. L. Robutti, R. C. Hoseney, and C. E. Wassom. 1974. Cereal Chem. 51:173-180.

Fig. 13. Reprinted, with permission, from: R. C. Hoseney, E. Varriano-Marston, and D. A. V. Dendy. 1981. Pages 77-144 in: Advances in Cereal Science and Technology, Vol. 4. Y. Pomeranz, ed. Am. Assoc. Cereal Chem., St. Paul, MN.

Fig. 15. Reprinted, with permission, from: K. Ghiasi, E. Varriano-Marston, and R. C. Hoseney. 1982. Cereal Chem. 59:262-265.

Fig. 17. Reprinted, with permission, from S. Takahashi and P. A. Seib. 1988. Cereal Chem. 65:474.

Figs. 18–20. Reprinted, with permission, from: K. Ghiasi, R. C. Hoseney, and E. Varriano-Marston. 1982. Cereal Chem. 59:258-262.

Fig. 22. Courtesy of Tom LuAllen, A. E. Staley Manufacturing Co., Decatur, IL.

CHAPTER 4

Figs. 1, 3, and 5. Reprinted, with permission, from: B. L. D'Appolonia, K. A. Gilles, E. M. Osman, and Y. Pomeranz. 1971. Pages 301-392 in: Wheat: Chemistry and Technology, 2nd ed. Y. Pomeranz, ed. Am. Assoc. Cereal Chem., St. Paul, MN.

Fig. 2. Courtesy of Jon Faubion.

Fig. 6. Reprinted, with permission, from: Y. Pomeranz. 1980. Cereal Foods World 25:659.

Fig. 10. Reprinted, with permission, from: G. Reed and J. A. Thorn. 1971. Pages 453-491 in: Wheat: Chemistry and Technology, 2nd ed. Y. Pomeranz, ed. Am. Assoc. Cereal Chem., St. Paul, MN.

Fig. 11. Reprinted, with permission, from: E. Graf. 1983. J. Am. Oil Chem. Soc. 60:1861.

Fig. 12. Adapted from: W. Grosch, G. Weber, and K. H. Fischer. 1977. Ann. Technol. Agric. 26:133.

CHAPTER 5

Fig. 2. Data courtesy of A. A. Abdelrahman.

Fig. 3. Reprinted, with permission, from: D. B. Sauer and R. Burroughs. 1980. Phytopathology 70:516-521.

Fig. 4. Reprinted, with permission, from: T. Labuza. 1984. Moisture Sorption: Practical Aspects of Isotherm Measurement and Use. Am. Assoc. Cereal Chem., St. Paul, MN.

Figs. 5 and 7. Reprinted, with permission, from: G. H. Foster. 1982. Pages 79-116 in: Storage of Cereal Grains and Their Products, 3rd ed. C. M. Christensen, ed. Am. Assoc. Cereal Chem., St. Paul, MN.

Fig. 6. Reprinted, with permission of the American Society of Agronomy, from: K. F. Finney, M. D. Shogren, R. C. Hoseney, L. C. Bolte, and E. G. Heyne. 1962. Agron. J. 54:244-247.

Figs. 8 and 9. Reprinted, with permission, from: G. H. Foster and J. Tuite. 1982. Pages 117-143 in: Storage of Cereal Grains and Their Products, 3rd ed. C. M. Christensen, ed. Am. Assoc. Cereal Chem., St. Paul, MN.

Fig. 10. Courtesy of Julie Liska.

Fig. 11. Courtesy of Kathy Zeleznak.

Fig. 12. Courtesy of John Pedersen.

CHAPTER 6

Fig. 1. Courtesy of Jeff Gwirtz.

Fig. 2. Courtesy of Steve Curran.

Fig. 3. Courtesy of Carter-Day Co., Minneapolis, MN.

Fig. 4. Courtesy of Forsbergs, Thief River Falls, MN.

Fig. 5. Reprinted, with permission, from: Stenvert and Kingswood, 1977. Cereal Chem. 54:627-637.

Figs. 6 and 7. Courtesy of Kathy Zeleznak.

Fig. 8. Courtesy of Buhler, Inc., Minneapolis, MN.

Fig. 9. Courtesy of Allis-Chalmers Corp., Milwaukee, WI.

CHAPTER 7

Fig. 2. Courtesy of Dorr-Oliver, Inc., Stamford, CT.

CHAPTER 8

Figs. 1 and 2. Reprinted, with permission, from: Borasio and Gariboldi. 1957. Illustrated Glossary of Rice Processing Machines. Food and Agriculture Organization of the United Nations, Rome.

Fig. 3. Courtesy of Satake Engineering Company.

Fig. 4. Reprinted, with permission, from: G. C. Witte, Jr. 1972. Pages 188-200 in: Rice: Chemistry and Technology, 1st ed. D. F. Houston, ed. Am. Assoc. Cereal Chem., St. Paul, MN.

Fig. 7. Reprinted, with permission, from: V. L. Youngs, D. M. Peterson, and C. M. Brown. 1982. Pages 49-105 in: Advances in Cereal Science and Technology, Vol. 5. Y. Pomeranz, ed. Am. Assoc. Cereal Chem., St. Paul, MN.

CHAPTER 9

Fig. 1. Courtesy of Kathy Zeleznak.

CHAPTER 10

Figs. 1 and 8. Reprinted, with permission, from: R. C. Hoseney, K. Zeleznak, and C. S. Lai. 1986. Cereal Chem. 63:285-286.

Fig. 2. Reprinted, with permission, from: R. C. Hoseney. 1979. J. Am. Oil Chem. Soc. 56:78A-80A.

Fig. 3. Reprinted, with permission, from: R. C. Hoseney and D. E. Rogers. 1990. Crit. Rev. Food Sci. Nutr. 29:73-93.

Fig. 4. Reprinted, with permission, from: R. J. Dimler. 1963. Baker's Dig. 37(1):52-57.

Figs. 5 and 7. Courtesy of G. Lookhart.

Fig. 6. Courtesy of D. D. Kasarda and H. P. Tao.

CHAPTER 11

Figs. 1 and 2. Reprinted, with permission, from: J. M. Faubion, P. C. Dreese, and K. C. Diehl. 1985. Pages 91-116 in: Rheology of Wheat Products. H. Faridi, ed. Am. Assoc. Cereal Chem., St. Paul, MN.

Figs. 3 and 4. Reprinted, with permission, from: K. R. Preston and R. C. Hoseney. 1991. Pages 13-19 in: The Extensigraph Handbook. V. F. Rasper and K. R. Preston, eds. Am. Assoc. Cereal Chem., St. Paul, MN.

Fig. 5. Reprinted, with permission, from: H. Faridi and V. F. Rasper. 1987. Pages 28-33 in: The Alveograph Handbook. H. Faridi and V. F. Rasper, eds. Am. Assoc. Cereal Chem., St. Paul, MN.

Fig. 6. Reprinted, with permission, from: B. Launay and J. Bure. 1970. Lebensm. Wiss. Technol. 3:57.

Fig. 7. Reprinted, with permission, from: E. B. Bagley and D. D. Christianson. 1986. Pages 27-36 in: Fundamentals of Dough Rheology. H. Faridi and J. M. Faubion, eds. Am. Assoc. Cereal Chem., St. Paul, MN.

Figs. 8-10. Reprinted, with permission, from: K. Shelke, J. M. Faubion, and R. C Hoseney. 1990. Cereal Chem. 67:575-580.

CHAPTER 12

Fig. 1. Reprinted, with permission, from: G. L. Rubenthaler and H. A. Faridi. Cereal Foods World. 28:627-629 (1983), and Baker's Dig. 55(6):19-22 (1981).

Figs. 2 and 25. Reprinted, with permission, from: K. Ghiasi, R. C. Hoseney, K. Zeleznak, and D. E. Rogers. 1984. Cereal Chem. 61:281-285.

Fig. 3. Courtesy of Grindstedt Products, Inc.

Fig. 8. Courtesy of K. Zeleznak.

Figs. 9, 14, and 20. Reprinted, with permission, from: R. C. Hoseney and P. A. Seib. 1978. Cereal Foods World. 23:362-367.

Fig. 10. Reprinted, with permission, from: R. C. Hoseney and P. L. Finney. 1974. Bakers Dig. 48(1):22-24, 26, 28, 66.

Fig. 11. Reprinted, with permission, from: E. D. Weak, R. C. Hoseney, P. A. Seib, and M. M. Baig. 1977. Cereal Chem. 54:794-802.

Figs. 12 and 13. Reprinted, with permission, from: R. C. Hoseney, K. H. Hsu, and R. C. Junge. 1979. Cereal Chem. 56:141-143.

Figs. 15 and 16. Reprinted, with permission, from: R. C. Junge, R. C. Hoseney, and E. Varriano-Marston. 1981. Cereal Chem. 58:338-342.

Fig. 17. Reprinted, with permission, from: K. F. Finney and M. D. Barmore. 1948. Cereal Chem. 25:291-312.

Fig. 18. Reprinted, with permission, from: M. D. Shogren and K. F. Finney. 1984. Cereal Chem. 61:418-423.

Fig. 19. Reprinted, with permission, from: M. D. Shogren and K. F. Finney. 1984. Cereal Chem. 61:179-181.

Figs. 21, 23, and 24. Reprinted, with permission, from He and Hoseney. 1991. Cereal Chem. 68:521-525.

Fig. 22. Reprinted, with permission, from He and Hoseney. 1992. Cereal Chem. 69:17-19.

Fig. 26. Reprinted, with permission, from Martin and Hoseney. 1991. Cereal Chem. 68:503-507.

FIGURE CREDITS / 371

CHAPTER 13

Fig. 1. Courtesy of Orion Research, Inc., Cambridge, MA.

Fig. 2. Reprinted, with permission, from: T. P. Kickline and J. F. Conn. 1970. Bakers Dig. 44(4):36-40.

Figs. 3, 5, 12, and 13. Courtesy of Werner and Pfleiderer, Stuttgart, Germany.

Fig. 4. Courtesy of Nabisco Brands, Inc., East Hanover, NJ.

Fig. 6. Courtesy of Franz Hass Machinery of America, Inc., Richmond, VA.

Fig. 7. Reprinted, with permission, from: A. M. Abboud and R. C. Hoseney. 1984. Cereal Chem. 61:34-37.

Fig. 8. Reprinted, with permission, from: K. Ghiasi, R. C. Hoseney, and E. Varriano-Marston. 1982. Cereal Chem. 60:58-61.

Fig. 9. Reprinted, with permission, from: L. C. Doescher and R. C. Hoseney. 1985. Cereal Chem. 62:263-266.

Fig. 10. Reprinted, with permission, from: L. P. Curley and R. C. Hoseney. 1984. Cereal Chem. 61:274-278.

Figs. 14 and 15. Reprinted, with permission, from: A. Pizzinatto and R. C. Hoseney. 1980. Cereal Chem. 57:249-252.

Figs. 16 and 17. Reprinted, with permission, from: K. Shelke, J. M. Faubion and R. C. Hoseney. 1990. Cereal Chem. 67:575-579.

CHAPTER 14

Fig. 1. Reprinted, with permission, from: H. Levine and L. Slade. 1990. Pages 157-330 in: Dough Rheology and Baked Product Texture. H. Faridi and J. Faubion, eds. Van Nostrand Reinhold, New York.

Fig. 2. Reprinted, with permission, from: L. Slade and H. Levine. 1991. Crit. Rev. Food Sci. Nutr. 30:115-360.

Figs. 3 and 4. Reprinted, with permission, from: J. W. Lawton. 1990. Cereal Chem. 69:351-355.

Fig. 5. Reprinted, with permission, from: K. Zeleznak and R. C. Hoseney. 1987. Cereal Chem. 64:121-124.

Fig. 6. Courtesy of D. Bath.

CHAPTER 15

Fig. 1. Reprinted, by permission, from: G. N. Irvine. 1971. Pages 777-796 in: Wheat: Chemistry and Technology, 2nd ed. Y. Pomeranz, ed. Am. Assoc. Cereal Chem., St. Paul, MN.

Fig. 3. Reprinted, with permission, from: H. J. Moss. 1973. J. Aust. Inst. Agric. Sci. 39:109-115. Pergamon Press Ltd.

Fig. 4. Reprinted, with permission, from: N. H. Oh, P. A. Seib, C. W. Deyoe, and A. B. Ward. 1983. Cereal Chem. 60:433-438.

Fig. 5. Courtesy of P. A. Seib.

Fig. 7. Courtesy of K. Rho.

CHAPTER 16

Fig. 1. Reprinted, with permission, from: V. L. Youngs, D. M. Peterson, and C. M. Brown. 1982. Pages 49-105 in: Advances in Cereal Science and Technology, Vol. 5. Y. Pomeranz, ed. Am. Assoc. Cereal Chem., St. Paul, MN.

Fig. 2. Courtesy of the Kellogg Company.

Figs. 3 and 4. Courtesy of Kathy Zeleznak.

Fig. 5. Courtesy of Wenger Manufacturing, Inc.

CHAPTER 17

Figs. 1-3. Reprinted, with permission, from: R. C. Hoseney, K. Zeleznak, and A. Abdelrahman. 1983. J. Cereal Sci. 1:43-52.

Figs. 6 and 7. Courtesy of Werner and Pfleiderer Corp., Stuttgart, West Germany.

Fig. 8. Courtesy of D. Bath.

CHAPTER 18

Fig. 1. Reprinted, with permission, from: J. R. Olson. 1983. Pages L1-L10 in: Proceedings of a Symposium on Particle Size Reduction in the Feed Industry. Kansas State University, Manhattan, KS.

Fig. 2. Reprinted, with permission, from: R. H. Leaver. 1988. The Pelleting Process. ABB Sprout-Bauer, Muncy, PA.

Fig. 3. Courtesy of Charles Stark.

Index

Pages followed by (f) are listed only for the figures they contain.

Acid-modified starch, 52-54
Activated double-bond compounds, 244
Adjunct cooking, 186-187
Adjuncts, 183-184
Aeration, 113-114
After-ripening, 168
Agglomeration, 141
Air classification, 137-139
Albumins, 68, 69, 70, 72, 73, 77, 78, 202
Aleurone, 69, 100, 141, 143
  structure, 6-7
  in various grains, 18, 19, 20, 21, 26
Alveograph, 219-221, 260
Amino acids, 65, 69, 197-200, 210
  composition in various cereals, 73, 74, 75, 76, 77, 78
Ammonium bicarbonate, 276
α-Amylase, 61, 94, 139, 140, 181, 183, 192, 233, 263, 331
β-Amylase, 36, 61, 94, 95, 183, 189, 192
Amylodextrin, 83
Amylograph, 45-49, 94
Amylopectin, 35-37, 38, 39, 40, 42, 44, 263
Amyloplast, 29
Amylose, 33-35, 38, 40, 43, 167
Ascorbic acid, 248
Ash, 136-137
Attenuation, 186
Attrition milling, 142, 143
Avenins, 77

Bacteria, 244, 248, 266, 291, 292
Bagels, 355, 356
Baking
  of bread, 254-258
  of cookies, 281-285, 286-290
  heat transfer, 256
Baking powder, 277-278, 280, 299
Barley
  for brewing, 177, 179
  flour, 143
  lipids, 90
  pearling, 161, 162, 173
  proteins, 68, 78
  starch, 29, 40, 42
  structure, 19
Batters, 223-226, 298, 299, 300, 302-303
Beer
  bottling, 191-192
  ingredients, 182-186
  light, 192
  production process, 186-191
Beeswing, 3-4
Benzoyl peroxide, 140
Birefringence, 31-32, 39, 40, 47, 51, 52, 140, 254
Biscuits, 278, 281, 303
Bran
  in kernel, 2(f), 69, 130, 131, 132, 135, 150, 338
  milling fraction, 6, 13, 125, 136, 148, 149
Bread
  crust, 315
  formula, 231-234
  processes, 234-237

  specialty, 264-267
  U.S. requirements, 230
  whole wheat, 265
Bread-making test, 269-270
Break system, 132-133
Breakfast cereals
  cooked, 335-336
  ready to eat, 209, 316, 336-343
Breather-type boxes, 173, 340
Brewing, 182-192
Browning of bread, 257, 258
Bumping, 338
Buns, 209
Butylated hydroxyanisole, 340
Butylated hydroxytoluene, 340
B-vitamins, 148, 166, 169

Cake finisher, 297
Cakes
  angel food, 301-302
  box mix, 296, 297
  high-ratio, 297
  ingredients, 298-301
  layer, 296-298
  pound, 302
Calcium hydroxide, 350, 351
Calcium propionate, 234
Carbon dioxide, 231, 244, 245, 248-251, 252, 256, 259, 276-278
Carotenoid pigments, 323, 325, 330
Caryopsis, 1, 2
Cell walls, 82, 85, 86

373

Cellulose, 5, 6, 18, 32, 81-82, 163
Cereal granules, 341
Chapati, 73, 351
Checking, 326
Chill haze, 191
Chlorine gas, 140, 299-300
Chorleywood procedure, 237, 241
Clathrates, 34, 58
Cleaning of grain, 125-129, 142
Color
    of corn, 13
    of wheat, 2-3
Complex viscosity, 309
Condensed tannins, 23, 184-185
Conditioning, 130-132
Continuous bread-making procedure, 236-237, 257
Cooker-extruder, 342, 343, 348
Cookies, 281-290
    baking, 286-290
    cutting-machine, 283-284
    dual-texture, 290
    flour quality, 285-286
    rotary-mold, 281-283
    sugar wafers, 284-285
    wire-cut, 284
Corn
    chips, 316, 351
    dry milling, 141-142
    drying, 111-112, 147
    oil, 93, 154-156
    proteins, 68, 69, 73-74, 148, 149, 150, 314
    snacks, 345-349
    starch, 40, 42, 49, 149-150, 209
    structure, 13-18
    syrups, 62, 184
    wet milling, 147-150
Corn chips, 316, 351
Corn curls, 348
Corn flakes, 336, 338
Corn-nuts, 347
Couscous, 324
Crackers, 244, 290-295
    saltines, 291-294, 316
    snack, 295
Cream of tartar, 278, 302
Creaming, 286, 296, 297
Crimping, 360
Cross cells, 5

Cross-linked kafirin, 75
Cross-linked zein, 74
Cross-linking, 54-57
Crust, 229, 254, 256, 257, 264, 267
    staling, 262, 263
Crystallinity, of starch, 30-31, 38, 39
Cysteine, 65

Danish doughs, 268
Decortication, 143
Denaturation, 67-68
Dextrins, 61
Dextrose, 60
Dextrose equivalent, 60, 61
Dicalcium phosphate, 280
Differential scanning calorimetry, 50-52, 286, 287, 312
Disulfide bond, 65, 73
Dormancy, 177-178
Dough
    bagel, 356
    biscuit, 303
    bubbles, 249, 250, 251
    danish, 268
    development, 238, 240-241
    effect of gluten on, 197, 201-202, 207, 244
    formation, 237-244
    frozen, 267-268
    mixing, 238-242
    nonwheat, 73, 351-352
    overmixing, 242-244
    refrigerated, 268, 303, 356
    rheology, 221-223, 245-248, 261, 267
    sweet, 268
    types, 73
    wheat, 70, 98, 208, 229
Dry rolling, 360
Drying
    of cereal grains, 108-113
    of pasta, 326, 328-329
Durum wheat, 10, 98, 207, 269, 321, 323, 324, 325
Dynamic rheometry, 215-217

Egg whites, 298, 301
Embryo, *see* Germ
Endocarp, 21, 22
Endosperm, 1, 6-13, 18, 69

opaque, 11, 14, 15, 25, 346
starchy, 6, 89
translucent, 11, 13, 15, 16, 25, 346
Enrichment, 141
Enthalpy, 51
Entoleter, 142
Enzymes, 93-99
    proteolytic, 96
Epicarp, 21
Equilibrium moisture content, 106
Ergot, 142
Ethanol, 231, 244, 248, 249, 256
Extensigraph, 217-219
Extrusion cooking
    of cereals, 342-343
    of feed, 365
    of pasta, 325-326
    of snacks, 316, 348

Falling number test, 94
Farina, 325, 335
Farinograph, 214
Feed efficiency, 360, 364
Feed formulation, 363
Fermentation, 190-191, 244-254
Ferulic acid, 85, 91, 92, 243
Flaking rolls, 338, 339
Flash dryers, 150
Flavonoid pigments, 328, 330
Flocculence, 186
Flour, 8
    cake, 299-300, 301
    clear, 136, 152
    cut-off, 135
    extraction, 136
    hard wheat, 135, 138, 238, 269, 275, 286
    malted, 140, 233
    patent, 135
    processing, 137-141
    quality, 209
        for bread, 229, 231, 252, 268-270
        for cookies, 285-286
        for noodles, 329-331
    rye, 142-143, 229, 252, 266
    self-rising, 141
    soft wheat, 135, 138, 269, 275, 286, 355

specialty, 209
straight-grade, 134(f), 135
strength, 208, 209, 218
treatments, 140–141
Free volume theory,
 309–310
Fructose, 60, 62
Fungi, 117–118
Furfural, 165, 173

Gas, in bread
 cells, 241, 245, 257, 261
 production, 233, 248–249
 retention, 249–251
Gelatinization, 31, 33, 40,
 42, 47, 54, 55, 351
Gelling, 49–50
Gels, starch, 56, 57, 58, 59
Germ, 1, 69, 116, 154
 composition, 78, 89, 141
 milling fraction, 132, 136,
  149, 150, 352
 structure, 6, 7
Germination, 110, 116,
 180–181
Glass transition, 204,
 307–318
 defintion, 308–309
 temperature, 307, 309,
  310, 312, 315
Gliadin, 201, 202, 207
Globulins, 68, 69, 70, 72,
 73, 77, 78, 202
β-Glucans, 7, 82, 85–86
Glucoamylase, 61
Glucofructosans, 86, 87
Glucono-δ-lactone, 280
Glucose, 59, 60, 61, 62, 86
 in starch structure, 31,
  33, 35, 36, 37, 38
 in yeast-leavened
  products, 248, 258, 268
Glucose isomerase, 62
Glumes, 1
Glutamine, 73, 78, 197, 199
Glutelin, 68, 69, 70, 77, 78,
 201
Gluten, 8, 71, 72, 270
 aggregates, 207
 in bread, 265
  dough, 244, 248, 250,
   251, 252–253, 258,
   260, 261
  staling, 263
 composition, 207–208
 in cookies, 282, 284, 286

in durum wheat, 323
genetics, 204–205
in glass transition, 312,
 314
hydrophobicity, 200–201
isolation, 197, 208–209
in noodles, 330, 331
nutritional quality, 210
physical properties,
 201–203
in rye flour, 266
structure, 198–199
synthesis, 206–207
thermal properties,
 203–204
uses, 209
wet milling, 150–153
Gluten-free bread, 254
Glutenin, 201, 202, 203,
 207, 208
Glycolipids, 89, 90, 91, 92,
 93
Glycoproteins, 68
Grain dust, 152
Grain sorghum, see
 Sorghum
Gravity table (separator),
 129, 160
Green malt, 181, 192
Grinding, 359–361
Grits, 110, 183, 336, 338
Groats, 21, 76, 170–173, 336
Gun puffing, 324

Hammer mill, 360
Helical inclusion
 compounds, 31, 34
Hemicelluloses, 7, 82–84,
 85, 86
High-fructose corn syrup,
 62
Hila, 347
Hops, 184–185
Hordein, 78
Hull, 1, 13, 18, 19, 78, 142
Humulone, 184
Husk, see Hull
Hyaline layer, see Nucellar
 epidermis
Hydroclones, 148, 149, 150
Hydrogen bonding, 199–200
Hydrophobic bonding, 200
Hysteresis, 106

Inoculum
 in baking, 291

in whiskey, 193
Insects, in stored grain,
 119–120
Integument, 22, 23, 25
Inulin, 86
Isoamylase, 37

Kafirin, 74, 75
Kan-sui, 328, 330
Kernel, see Caryopsis
Keyholing, 233, 261
Kilning, 181–182

Lactobacilli, 148
Lagering, 191
Lauter tub, 189
Leavening, chemical,
 276–281
Lemma, 15(f), 18, 19, 163
Lignin, 18, 163
β-Limit dextrin, 37, 94, 95
Lipase, 97, 166, 170
Lipids, 32, 87–93, 197,
 200–201, 252
 bound and free, 89
 nonpolar, 89, 90, 91, 92,
  93
 polar, 89
Lipoxygenase, 98–99, 244,
 325
Loss modulus, 217, 224
Lupulone, 184
Lysine, 69, 72, 75, 77, 78,
 198, 210

Macaroni, 321
Maize, see Corn
Malt, 181, 182, 183
Malting, 177, 179–182
Maltose, 36, 37, 60, 61, 94,
 95, 183
Maltose value, 95
Maltotriose, 37, 60
Masa, 349–351
Mashing, 187–189, 193
Meat analog, 209
Mesocarp, 21, 23(f)
Microflora, 117–118
Micronizing, 363
Milk and milk replacer,
 231, 233, 298
Millet, see Pearl millet
Milling
 attrition, 141–142
 changes caused by, 137
 products, 135–137

376 / INDEX

steps in the process, 125–135
wet, 125, 147–156
Minerals, 99–100
Mixing, 238–241
multistage, 296
of pasta, 325
single-stage, 296, 297
Mixing time, 241–242
Mixograph, 214
Moisture, in grain
equilibrium, 106–107
in storage, 105–108, 114, 120
uptake, 130, 131
Molding, 249, 254, 257
Molds, 104, 107–108, 116
Monocalcium phosphate, 278
Monogastric animals, 360
Monoglycerides, 233–234, 343
Mutants, 40, 42, 43, 69
Mycotoxins, 117–119, 166

Natural rubber, 222
Neutralization value, 278
Newtonian viscosity, 214
Nixtamal, 350
Noodles, 327–333
cutting, 331
flour for, 329–331
ramen, 329
sheeting, 331
udon, 329
Nucellar epidermis, 1, 6
Nutritional value, of cereals, 69–73, 76, 125, 169, 210

Oats
lipids, 91
milling, 170–173
proteins, 76–77
rolled, 172, 336, 340
starch, 42, 46(f)
structure, 21
Oil production, 153–156
Oils, essential, 184
Oligosaccharides, 86–87
Opaqueness, 11, 13, 56, 57, 58
Oryzenin, 77
Oscillatory rheometry, 224, 226
Oven puffing, 342

Oven-spring, 256, 259
Overmixing, 242–244
Oxidants, 140
chemical, 231, 233, 237, 247, 248, 267
fast-acting, 244, 248
Oxidative gelation, 84–85

Paddy, 18, 159, 160, 163, 168, 169
Paddy machine, 160, 172
Palea, 15(f), 18, 19, 163
Parboiling, 169
Pasta, 207, 321–326
production process, 325–326
Paste, starch, 50, 56
Pasteurization, of beer
Pasting, 47, 55–56
Pearl millet
flour, 143
lipids, 92–93
proteins, 75–76
snacks, 347–348
starch, 42
structure, 25
Pearler, 162
Pearling, 143, 161, 162, 173
Pellagra, 74, 76
Pelleting, 361–363
Pentosans, 82–84, 85, 163, 251, 252, 266
Pericarp, 1, 3, 5, 69, 346, 351
Peroxidase, 99
Pet foods, 209
Phospholipids, 33, 89, 90, 91, 92, 93
Phytase, 97–98
Pigments
carotenoid, 323, 325, 330
flavonoids, 328, 330
Pitching, 190
Plasticizing 312, 313, 339
Polymers, definition, 307
Polyphenoloxidase, 328, 331
Polysaccharides, nonstarchy, 81–86
Popcorn, 345–347
Popped pearl millet, 347–348
Popped sorghum, 347–348
Popping feed, 363–364
Potassium bicarbonate, 276
Preferment bread-making system, 236–237, 247, 248

Pretzels, 353–355
soft, 355
Prolamins, 68, 69, 70, 73, 74, 75, 76, 77, 78, 201
Proline, 198, 199
Proof
in baking, 235, 249, 303
of whiskey, 193
Proofing, 254, 267
Propylene glycol monostearate, 296, 299
Proteases, 96
Protein
bodies, 8, 13, 19, 24, 25, 206, 346
classification, 68
content in various cereals, 5, 7, 69–70, 71–72, 73, 76, 77
in dough, 208, 242
and flour quality, 252–253, 269–270
storage, of wheat, *see* Gluten
structure, 65–68
Protein-starch bond, 10, 11, 14, 16
Proteolytic enzymes, 96
Puff pastry, 268
Puffed cereal, 341–343
Puffing methods, 341
Pullulanase, 37, 61
Purifier, 135

Quick-cooking cereals
oats, 172–173
rice, 166, 168–169

Ramen noodles, 329
Rancidity, 125, 141, 143, 169, 172, 336, 340
Ready-to-eat cereals, 336–343
Reducing agents, 242
Respiration, 115–116
Retarder, 268
Retrogradation, 35, 49–50, 56, 58, 262–263
Rheology, 213–227
Rice
aging, 168
bran, 161, 162, 163, 165–166
brown, 18, 160, 161, 165
drying, 112–113
enrichment, 169–170

head, 163, 166, 169
hulls, 159, 160, 163–165
lipids, 92
milling, 162–163
oil, 154, 166
parboiling, 169
polish, 163, 165–166
processing, 157–170
proteins, 68, 77, 167
quality, 166–167
quick-cooking, 166, 168
rough, 18, 159, 160
starch, 153, 167
structure, 18–19
Roasting feed, 364
Rodents, in stored grain, 121
Rolled oats, 336, 338, 340
Roller milling, 132–135, 142
Roti, 352
Ruminant animals, 360
Rye
  bread, 142–143, 229, 266
  flour, 143, 252, 266
  lipids, 91
  milling, 142–143
  proteins, 68, 73, 77
  starch, 29, 40, 41(f), 42
  structure, 20

Salt, 231, 242, 276, 280, 294
Saltine crackers, 291–294
Scourer, 128–129
Seed coat, 1, 6, 18, 20, 22, 26
Semolina, 321, 323, 324
Separators
  disk, 128, 163, 171
  gravity, 160
  liquid cyclone, 148
  magnetic, 126
  milling, 126, 128
Setback, 49
Sheer thinning, 48, 55, 58
Shortening, 232, 263, 268, 281, 282, 286, 288, 297, 298
Shorts, 134(f), 135
Short-time baking systems, 237, 245
Shredded biscuits, 340
Shriveling, 26
Sick wheat, 116, 117(f)
Sifting system, 133, 135
Silica, 18, 163, 165
Sludge, 83

Snack crackers, 295
Sodium acid pyrophosphates, 278–279, 303
Sodium aluminum phosphate, 279–280
Sodium aluminum sulfate, 280
Sodium bicarbonate, 276, 277, 285, 286, 292, 355
Sodium hydroxide, 354, 355
Sorghum
  flour, 143
  lipids, 93
  proteins, 68, 74–75
  snacks, 347–348
  starch, 40, 42, 150
  structure, 21–24
Sourdough bread, 244, 266–267
Spaghetti, 321, 326, 332
Specialty feeds, 365
Sponge-and-dough process, 235–236, 248, 291, 292
Squeegee, 83
Staling, 262–263
Starch
  in bread, 261, 263
  in cakes, 296, 302
  composition, 32–38, 40–44
  in cookies, 287–288
  crystallinity, 30–31, 38, 39
  damaged, 139, 140, 275, 303
  fractionation, 38
  glass transition of, 314–315
  granules, 29–32, 38–39, 45, 46, 47, 49, 346
  lipids, 32, 89–90
  modified, 52–59
  in noodles, 330
  solubility, 47, 48, 56, 58
  in various cereals, 8, 10, 13, 15, 19, 20, 21, 24, 25, 40–44
  waxy, 38, 43
Starch-water-heat, effect of, 44–52
Steam flaking, 365
Steam rolling, 360
Steeping, 147–148, 179, 180
  solids from, 148
Storage

of beer, 191
of grain
  changes during, 114–117
  damage during, 117–121
  types, 104–105
Storage modulus, 217, 224
Storage proteins, 69, 70
Straight-dough process, 234–235, 248–249
Strain, 214, 216, 217
Stress, 214, 216
Stress cracking, 110
Substituted starches, 57–58
Sucrose, 60, 62
  in baking, 248, 249, 258, 268, 289, 290, 298
  in plants, 29, 30
Sugar
  in baking of soft wheat products, 287, 288, 296, 297–298, 300–301
  in bread formulas, 232–233, 265, 268
  in cereal grains, 7, 86–87, crystallization, 289, 290
  in fermentation, 245, 248–249, 250
  glass transition, 317–318
  recrystallization, 290
Sugar coatings, 343
Sulfur dioxide, 147, 148, 152, 182
Surfactants, 231, 233–234, 250, 254, 263, 299
Sweeteners, 59–62
Syneresis, 50
Synthetic nuts, 352–353

Tailings, 83
Tan δ, 217, 224
Tempering, 130–132
Testa, 22
Thermal analyzer, 311–312
Tortilla, 351
Trichomes, 21, 170
Triticale
  proteins, 73, 77
  starch, 40
  structure, 26
Tube cells, 5, 6, 22

Udon noodles, 329
Uniaxial compression, 221

Vitamins, 99, 100, 167
 B-vitamins, 7, 148, 166, 169
 E-vitamins, 7, 89
Vitreousness, 11, 13, 24
Vitrification, 308, 309
Vulcanized rubber, 122, 223

Water, 231, 312, 313, 352
 in baking, 276, 282, 290, 294
 for brewing, 185
 uptake, 130
Waxy starch, 38, 43
Wet milling, 125, 147–156

Wheat
 bread, *see* Bread
 gluten, *see* Gluten
 hard, 7, 8, 10, 11, 130
 lipids, 89
 oil, 154
 proteins, 68, 69, 70–73
 soft, 7, 8, 10, 11, 130, 135, 139
 starch, 29, 40, 42, 150–153, 208
 structure, 2–13
 sugars, 87
 wet milling, 150–153
Wheat flakes, 338–339

Whiskey, 193
Winterization, 156
Wort, 189–190

Yeast
 in bread, 229, 231, 256, 258, 266, 267, 268
 in brewing, 185–186
 fermentation process, 244–245, 247, 248, 249
 in soft wheat products, 276, 291, 292

Zein, 73, 74, 75
Zein bodies, 13, 16